The Rise & Fall of Development Theory

The Rise & Fall of Development Theory

COLIN LEYS

Professor of Political Studies
Queen's University
Kingston, Ontario

EAEP
NAIROBI

INDIANA UNIVERSITY PRESS
BLOOMINGTON & INDIANAPOLIS

JAMES CURREY
LONDON

James Currey Ltd
54b Thornhill Square
Islington
London N1 1BE, England

Indiana University Press
601 North Morton Street
Bloomington, Indiana 47404, USA

East African Educational Publishers
P.O. Box 45314
Nairobi
Kenya

This collection first published 1996

1 2 3 5 6 00 99 98 97 96

British Library Cataloguing in Publication Data
Leys, Colin
Rise and Fall of Development Theory
I. Title
306.3091724

ISBN 0-85255-350-1 (James Currey Paper)
ISBN 0-85255-359-5 (James Currey Cloth)

A CIP catalog record for this book is available from the Library of Congress

ISBN 0-253-21016-X (Indiana University Press Paper)
ISBN 0-253-33083-1 (Indiana University Press Cloth)

Typeset in 10/12pt Plantin by Nicholas Hardyman
Printed and Bound in Britain by Villiers Publications, London N3

Contents

Preface

It is more than thirty years since 'development' emerged as the focus of heated and — for a time at least — intellectually stimulating theoretical debates, and more than thirty years since most of sub-Saharan Africa became independent; but as the century nears its end both development theory and Africa are in deep trouble. In the first and last essays in this collection I argue that one of the main reasons is largely the same in both cases: the end of the regulated system of national economies formalized at Bretton Woods, which underlay the whole idea of 'development' as it was conceived from the 1950s onwards. If this is correct, a major challenge for development theory is to confront the problem of how we can now resubordinate 'the market' to a new system of international and national regulation, and, as a necessary corollary of this, to clarify the political assumptions and commitments on which the theory rests.

Political commitments were what once gave debates about development theory their energy, and what has demobilized development theory is a retreat from politics in face of the seemingly irresistible success of capital in subordinating everything to the arbitration of 'global market forces'. But a broader, more historical view of development shows that no such victory is ever really irresistible. On the contrary, the present 'impasse in development theory' and the unfolding tragedy in Africa are not due to the working out of an inexorable law of economics but, to a significant extent, to politically motivated policy decisions (setting capital free to pursue profit wherever it wishes and on whatever terms it can impose), rationalized by a particular brand of development theory (neo-liberalism) which assigns all initiative to 'the market' (i.e. to capital). If development theory is to be useful and interesting again it must focus carefully on these decisions and consider ways and means to re-create a world in which it is once again possible to pursue social goals through the collective efforts of the societies and communities to which people belong.

This in turn means going back, although in a self-critical and open-minded spirit, to the classical agenda of political economy. The world has never been as fully exposed to the operation of 'market forces' as it is today, so, whatever the shortcomings of past efforts to understand these forces and their social and political consequences, a first task must be to understand them better now. This is the main theme of the papers collected here.

In staying within this range of issues I am very conscious of continuing to neglect a number of other closely connected and important questions that recent research and practice have put on the development agenda, and especially two: the condition and needs of women and the indispensable contribution of women to the solution of development problems, and the philosophical, economic and social dilemmas that the limits of the physical environment pose for development, however it is conceived. However, I believe that the rapidly accumulating wealth of information and analysis on these issues can only make its full impact on development when it is integrated with a reinvigorated political economy.

Two sub-themes within the chapters that follow deserve a brief mention. One that recurs in the first part of the book is the politics of American political science. The USA is a superpower with imperial interests, a markedly military history and a tradition of intolerance towards the left. These things inevitably influence American political science: the congruence beween the standpoint of so much mainstream American political science and US foreign policy (including the neo-liberal reconstruction of the world order in the 1980s) is a significant fact to which attention should be drawn.

A second sub-theme, which runs through Part II, is the role of domestic capitalists in African development. Although I think that a reconstituted international regime of regulated trade and capital flows is now a precondition for renewed development in most sub-Saharan African countries, another precondition is a radical rethinking of the framework (market and non-market institutions) in which capital can be expected to be accumulated in African conditions, and the emergence of a stratum of people capable of directing the accumulation of capital and interested in the political conditions which make this possible. These need not constitute a ruling class, but the function they perform is indispensable and needs to be much better understood.

Chapters 1, 4 and 9 are new (although part of Chapter 1 and parts of Chapter 9 draw on an article entitled 'Confronting the African Tragedy' published in *New Left Review* 204, 1994). Chapter 1 develops the general argument already sketched above. Chapters 2, 3 and 4 reflect critically on dependency theory, modernization theory and the 'new institutionalism'/new political economy respectively. Chapter 2 is included partly because it has been fairly widely referred to, but also because it can still be a useful reminder of the unavoidable trade-offs between consistency and political relevance that are involved in any worthwhile theoretical initiative. Chapter 3, on the earlier work of Samuel Huntington, is there to remind the reader of the real nature of the most widely used text of its day in the American political science of development and as Huntington has been at pains to remind us with his 1993 article, 'The Clash of Civilizations?' [*Foreign Affairs* 72/3, pp. 22–49], modernization theory, and the racially tinged imperialism that informs Huntington's version of it, are by no means things of the past. Chapter 4 offers a critique of a contemporary current in US political science that I think serves, however unintentionally, to rationalize an uncritical view of the 'market' (capital) and its effects on the 'periphery'.

Chapter 5, on African political economy at the beginning of the 1980s, has been included mainly because those years were such a watershed in Africa. Although the article registered the gathering pessimism about both African capitalism and African socialism, it contained no hint of the impending ascendancy of the neo-liberal doctrines advocated by the Berg Report on *Accelerated Development in Sub-Saharan Africa,* which appeared in the year the article was written (even though it made reference to it), and it was far too optimistic about the progressive potential of the new social forces produced by development in the 1960s and 1970s. Chapters 6, 7 and 8, written in 1987, 1991 and 1993 respectively, reflect the altered climate of those years. Chapter 6, on the crisis of petty commodity production in African agriculture, is included because the original article is not always easily accessible, and because its central point seems to have become, if anything, more significant with the passage of time. Chapter 7 reviews the so-called 'Kenya debate' that was prompted by an article of mine published in 1978, and leads into a more general discussion of the role of African capitalists in African development in Chapter 8. Chapter 9 brings together the two parts of the book, arguing that both the current African tragedy and wider considerations point in the same theoretical and political direction.

Although it is surely important to be capable of changing one's mind, and I would not now say everything contained in the earlier papers collected here (or at least not in just the same terms), I still broadly agree with the main lines of argument in them. In any case I have altered very little apart from some minor shortening of Chapter 5 and a few places where I have changed a word or phrase to make the text read less awkwardly or with less asperity.

I would like to thank James Currey for his enthusiasm and patience over two decades; Lynn Taylor and Nicholas Hardyman for their highly professional support; and the original publishers for permission to use the articles reprinted here. I am also very grateful to Manfred Bienefeld, Michael Lebowitz, Roger Leys, John Saul, Nancy Leys Stepan and Donald Swartz for invaluable comments on various drafts, and especially to Agnes Callamard, who went to exceptional trouble to give me detailed, penetrating criticisms and many helpful suggestions. None of these friends is, of course, in any way responsible for the results.

I

Development Theory

1

The Rise & Fall of Development Theory

> To say one more word about preaching what the world ought to be like, philosophy arrives always too late for that. As *thought* of the world it appears at a time when actuality has completed its developmental process and is finished...When philosophy paints its gray in gray, a form of life has become old, and this gray in gray cannot rejuvenate it, only understand it. The owl of Minerva begins its flight only when dusk is falling.[1]

> Only with this universal development of productive forces is a *universal* intercourse between men established, which produces in all nations simultaneously the phenomenon of the 'propertyless' mass (universal competition) and makes each nation dependent on the revolutions of the others, and finally has put *world-historical*, empirically universal individuals in place of local ones...How otherwise...does it happen that trade...rules the whole world through the relation of supply and demand — a relation which, as an English economist says, hovers over the earth like the fate of the ancients, and with invisible hand allots fortune and misfortune to men, sets up empires and overthrows empires, causes nations to rise and to disappear...?[2]

As the dusk gradually falls on the twentieth century, the owl of Minerva has a lot on its agenda; with so many illusions to clear up, those for which 'development theory' is responsible might be thought to have a rather low priority. Yet what is at stake is — or rather was, since the practical ambitions of 'development theory' have been progressively reduced over the years — nothing less than whether human beings can act, collectively, to improve their lot, or whether they must once again accept that it is ineluctably determined by forces — nowadays 'world market forces' — over which they have, in general, little or no control (and least of all those who need it most). Unfortunately, in spite of the importance of the question, 'development theory' has returned only partial and conflicting answers to it.

[1] G.W.F. Hegel, 'Philosophy of Right and Law', in Carl J. Friedrich (ed.), *The Philosophy of Hegel* (New York: Random House, 1953), p. 227.
[2] Karl Marx and Friedrich Engels, 'The German Ideology', in Robert Tucker (ed.), *The Marx–Engels Reader* (New York: W.W. Norton, second edition 1978), pp. 161–62.

The first theories of development

To see why, it is useful to begin with an owl's eye view of human history over the past 10,000 years or so since settled agriculture first began to replace hunting and gathering. Agriculture required, but also made possible, an increased specialization of labour, and the development of state apparatuses capable of organizing the defence of cultivated land against outside aggressors and of assuring stability in the increasingly complex social and economic relations on which an agricultural society gradually came to depend. With the establishment of agriculture, then, the process of social evolution greatly accelerated relative to that which had occurred during the preceding 1.8 million years of human life on earth; but for a long time (we may suppose) the process was still sufficiently gradual, and sufficiently precarious at any given place and time, for it not to be felt as an acceleration by the people living through it.

The advent of capitalism in the fifteenth and sixteenth centuries, however, and above all the advent of industrial capitalism in the late eighteenth century, forced the fact of human economic, social, political and cultural development on people's attention. Various thinkers, from Condorcet to Kant, began to conceive of a 'universal history' which would disclose the cumulative pattern and meaning of it all, and its ultimate destination;[3] but the decisive innovators were, of course, Hegel and Marx. Hegel saw world history as a process of development, 'a progression to the better'; but since in his view world history took place 'in the field of the spirit' this progression was only an 'actualization' of the evolving self-understanding of the 'world spirit', in which partial conceptions of the idea of freedom were successively superseded by more and more complete conceptions, through a dialectical process of contradictions and their transcendence.[4] Marx at first followed Hegel, but then jettisoned his metaphysics while retaining the idea of history as progress, seeing it instead as a series of modes of production, each disclosing a higher level than before of the uniquely human capacity for self-realization, and succeeding each other through the working out of contradictions in their inner dynamics.

What makes Hegel and Marx true originators of development theory is that they recognized that it was the sudden acceleration in the rate of change that the establishment of capitalist production and bourgeois society had generated that made it necessary and possible to think of history in this way. Bourgeois society had to be understood historically if it was to be made rational (Hegel's idea), or superseded (Marx's); but this understanding, both of them realized, in which capitalist society was seen as the

[3] See the neat résumé of this in Francis Fukuyama, *The End of History and the Last Man* (New York: The Free Press, 1992), Chapter 5.

[4] G.W.F. Hegel, 'The Philosophy of History', in Friedrich (ed.), *The Philosophy of Hegel*, op. cit., p. 11 and *passim*; he concludes (p. 23): 'World history presents therefore the stages of the principle whose meaning is the consciousness of freedom.'

outcome of an evolutionary process stretching back into the mists of time, should also make possible an adequate understanding of earlier societies; as Marx put it, 'Human anatomy contains a key to the anatomy of the ape.'[5] Between them they inspired a vast subsequent output of theory-inspired historiography and historically based social science concerned with understanding the evolution of human life on earth as a structured totality.

The emergence of 'development theory'

But this tradition of thought about development is not what most people have meant by the term 'development theory', which emerged in the 1950s to deal with a far narrower issue: namely, how the economies of the colonies of Britain, France, Portugal and other European powers, colonies comprising some 28% of the world's population, might be transformed and made more productive as decolonization approached, in the context of the still 'semi-colonial' condition of the former colonies of Latin America (accounting for a further 7%).[6] Understanding this unprecedented event, and gearing policy to these aims, unquestionably called for new theoretical work. But it is striking how little of this work drew on, or even related itself to, the existing body of theory about development that had been prompted by the original advent of capitalism itself.

There were three main reasons for this. First of all, the new 'development theory' had a very strong practical orientation: its aim was to provide grounds for immediate action. Even academic theorists — as opposed to those directly working for development agencies of one sort or another — were drawn to the field by a desire to do something for the peoples of the ex-colonies, and had an even higher degree of conscious commitment to intervention than is usual in most other branches of social science. This militated against philosophical dispassion and reflective self-criticism.

Secondly, the 'new nations' were a prime stake in the Cold War, so that theories of their development were unavoidably contaminated by this. Of course, most development theorists saw their work as science, not propaganda: few were interested in following the example of W.W. Rostow by

[5] Karl Marx, 'Introduction' to the *Grundrisse*, in Tucker (ed.), *The Marx–Engels Reader*, op. cit., p. 241: 'Bourgeois society is the most developed and the most complex historic organization of production. The categories which express its relations, the comprehension of its structure, thereby allow insights into the structure and the relations of production of all the vanished social formations out of whose ruins and elements it built itself up...Human anatomy contains a key to the anatomy of the ape...The bourgeois economy thus supplies the key to the ancient, etc.'

[6] Estimated from data in Colin McEvedy and Richard Jones, *Atlas of World Population History* (New York: Facts on File, 1978). Fifty years later, population growth had produced a situation where the 'developing countries' (including China) accounted for almost four-fifths of the world's population.

subtitling any of their works 'A Non-Communist Manifesto'.[7] But, whereas the early theorists of rising capitalism thought it essential to locate it in a broad conception of history, most Western theorists of development in the post-war years (and most of them were Westerners) avoided doing so because it meant, unavoidably, taking seriously the work of Marx, which at the height of the Cold War was not merely considered unscientific, but in the USA could easily cost you your job.[8] As a result 'development studies' tended to be conducted, at least until the mid-1960s, as if they had no significant historical or philosophical roots or presuppositions; and while 'development theorists' were usually glad to affirm their strong normative reasons for being concerned with development, they rarely acknowledged the extent to which their thinking reflected their own political commitments.[9]

A third crucial conditioning factor in the birth of 'development theory' was the Bretton Woods financial and trading regime. These arrangements were designed to permit national governments to manage their economies so as to maximize growth and employment. Capital was not allowed to cross frontiers without government approval, which permitted governments to determine domestic interest rates, fix the exchange rate of the national currency, and tax and spend as they saw fit to secure national economic objectives. National economic planning was seen as a natural extension of this thinking, as were domestic and international arrangements to stabilize

[7] W.W. Rostow, *The Stages of Economic Growth: A Non-Communist Manifesto* (Cambridge: Cambridge University Press, 1960).

[8] The degree to which critical social science was systematically rooted out in the USA from 1948 onwards is apt to be forgotten, both by Americans, for whom its consequences have come to be taken as natural, and by foreigners who never experienced it. The consensus of historians is painfully clear that not only did several hundred university and college professors and high school teachers lose their jobs, while many more were prevented from getting appointments, because of their political views (real or just alleged); but also the complicity of most academic administrators, the systematic use of informers and the weakness of the Association of American University Professors (AAUP) combined with the general jingoism of the time to eliminate virtually any teaching or research that questioned American state policy; 'open criticism of the political status quo disappeared...teachers played it safe, pruning their syllabi and avoiding controversial topics...The academy's enforcement of McCarthyism...silenced an entire generation of radical intellectuals and snuffed out all meaningful opposition to the official version of the Cold War'(Ellen W. Schrecker, *No Ivory Tower: McCarthyism and the Universities* (New York: Oxford University Press, 1986), pp. 339 and 341). See also David Caute, *The Great Fear: The Anti-Communist Purge Under Truman and Eisenhower* (New York: Simon and Schuster, 1978), pp. 403–30, and Lionel S. Lewis, *The Cold War on Campus* (New Brunswick: Transaction Books, 1988).

[9] A striking illustration of this can be found in Gabriel Almond's contribution to the review of late modernization theory edited by Myron Wiener and Samuel P. Huntington, *Understanding Political Development* (Boston: Little, Brown, 1987), in which he represents the modernization school as 'objective' ('governed by professional criteria of evidence and inference') and its critics as mere propagandists (pp. 444–68 and especially 450 ff.).

commodity prices.[10] It is not a great oversimplification to say that 'development theory' was originally just theory about the best way for colonial, and then ex-colonial, states to accelerate national economic growth in this international environment. The goal of development was growth; the agent of development was the state and the means of development were these macroeconomic policy instruments. These were taken-for-granted presuppositions of 'development theory' as it evolved from the 1950s onwards.

For over ten years (i.e. from 1955 to the late 1960s) 'development theory' so conceived progressed with only modest excitement. Then, partly due to disappointment with the results of policies based on 'development theory' (especially in Latin America and India), and partly to the general reaction of the 1960s against all 'official' values and ideas, the theoretical temperature rose. The ahistorical, unself-critical and politically partisan nature of 'development theory' was put in question by critics on the 'left'; and one way to understand the heady debates that followed throughout most of the 1970s is as a struggle between those who tried to keep 'development theory' within its original parameters, and critics who were trying to extend them and place the issues back into the framework of the historically orientated and ethical tradition of general development theory founded by Hegel and Marx.

The full implications of doing this were, however, obscured for a long time by the fact that most of the critics also subscribed to a very practical, short-term, state-orientated conception of development (and in many cases were also influenced by Cold War partisanship).[11] But the work of finally demonstrating the limitations of mainstream 'development theory' was not left to be accomplished by criticism alone. By the mid-1980s the real world on which 'development theory' had been premissed had also disappeared. Above all, national and international controls over capital movements had been removed, drastically curtailing the power of any state wishing to promote national development, while the international 'development community' threw itself into the task of strengthening 'market forces' (i.e. capital) at the expense of states everywhere, but especially in the Third World. As a result most states could no longer be the prime movers of development that 'development theory' had hitherto always presupposed, and none of the alternative candidates (such as 'social movements' or 'communities') proposed by 'development theorists' as the field unravelled were very convincing.

But this is to anticipate. The next section seeks to make more persuasive

[10] As is well known, the US position in the negotiations leading to the Bretton Woods agreement was more *laissez-faire*. By the mid-1960s, however, national economic planning had become conventional economic wisdom in Western Europe, and not just for 'developing countries'.

[11] e.g. the Maoism of Pierre-Phillippe Rey, in *Les Alliances de classes* (Paris: Maspero, 1973) and the Guevarism of Debray, in *Revolution in the Revolution? Armed Struggle and Political Struggle in Latin America* (New York: Monthly Review Press, 1967).

the general characterization of 'development theory' given above by recalling briefly how the contents of some of the main currents of development theory were determined by who produced them, under what conditions and why.

Development theories: science and discourses[12]

The first formulations of development theory were the work of economists, all strongly influenced by the ideas of Keynes and the wartime and post-war practices of state intervention in the economy, including the perceived success of the Marshall Plan, which was in many ways a model for later ideas about 'aid'. They shared the broadly social-democratic ethos of the period, including its commitment to planning and its conviction that economic problems would yield to the actions of benevolent states endowed with sufficient supplies of capital and armed with good economic analysis. They produced what P.W. Preston has aptly called development theory's 'positivist orthodoxy'.[13] They wrote development plans for both newly independent countries and the not yet independent colonies of Africa, based on the idea of raising rural productivity and transferring underutilized labour out of agriculture into industry.[14]

By the end of the 1950s, however, the original optimism that this approach would yield rapid results had begun to evaporate, and the limitations of development economics as a theory of development were beginning to be exposed. The failure of the Indian economy, in particular, to respond rapidly to this approach was attributed in part to the 'softness' of the Indian state, which seemed to lack the capacity to live up to the social-democratic ideal of a rational, firmly benevolent enforcer of the national interest and impose the necessary discipline on everyone from businessmen and landlords to small peasants.[15] But this famous judgement, coming from

[12] An outstanding recent contribution to the study of development theory as discourse is David B. Moore and Gerald G. Schmitz (eds), *Debating Development Discourse: Institutional and Popular Perspectives* (London: Macmillan, 1994); see especially the penetrating and rich opening chapter by Moore, 'Development Discourse as Hegemony: Towards an Ideological History — 1945–1995', pp. 1–29. Moore's position is at many points similar to, and has also influenced, my own. Moore focuses primarily on development theory's role in the struggle for international hegemony. While this dimension is politically crucial it seems to me likely that on further investigation the professional and organizational dimensions of development discourses (some of the elements of which are outlined in his chapter) will acquire more salience.

[13] P.W. Preston, *Theories of Development* (London: Routledge, 1982), Chapter 3, referring to the work of Harrod and Domar and their successors, 'exported' to the soon-to-be ex-colonies.

[14] The key text here was Arthur Lewis, 'Economic Development with Unlimited Supplies of Labour', *Manchester School* 22/2, May 1954, pp. 139–91.

[15] Gunnar Myrdal, *Asian Drama: An Enquiry Into the Poverty of Nations* (New York: The Twentieth Century Fund, 1968), Vol. I, p. 66. See also Chapter 18, sections 13–14.

Gunnar Myrdal and his associates (notably Paul Streeten, and later Dudley Seers), representing the 'left' (and most historically and sociologically sensitive) wing of development economics, signalled the existence of complex problems which lay beyond the conceptual and empirical scope of mainstream — i.e. neo-classical — economics.[16] Marx had long ago grasped that states were, as he put it, but the 'official résumés' of civil society. In the first phase of development economics this had been forgotten. What was it about these societies that made them unresponsive to the 'positivist orthodoxy'?

'Modernization theory' was an American response to this question. It was constructed by sociologists and political scientists involved in the rapidly expanding research and teaching programmes established by the US government to equip the country with the regional expertise it needed to exercise its new role as a superpower. These experts none the less found themselves largely excluded from policy-making roles in the US Agency for International Development (USAID) or the World Bank, the two most important aid agencies in the world, both headquartered in Washington; and modernization theory can be understood in part as their explanation of why the plans of development economists who monopolized these organizations so seldom worked. They believed that in the transition from 'traditional' to 'modern' forms of social organization, already completed in the industrialized West, the complex interactions between social change and economic development, mediated by politics, could be traced with some precision, using 'structural-functional analysis' and a typology of social structures derived from Weber by Talcott Parsons. This belief had a fervent quality that survives even in this retrospective summary by one of its most prominent practitioners:

> Structural-functional analysis...is concerned with the connections between norms, structures, and behaviors, the first in terms of values and interests, the second in terms of roles, role networks, and classes, and the third in terms of motivation and perception. An equilibrium model, its political focus is on the stability of complementary and mutually reinforcing relations between each of the dimensions. The three together constitute a system. Change in any one will effect alterations in the others...Within these abstract formulations it is possible to organise highly complex masses of data around a host of subsidiary variables...There is a clarity in this framework and, if it is used with a certain

[16] Myrdal, who had served for ten years as head of the secretariat of the Economic Commission for Europe, was intensely aware of the limitations of a purely economistic approach to development problems, i.e. one restricted to what can be analysed within the framework of neo-classical economics. He was also keenly aware of the Cold War context, and reading *Asian Drama* makes a painful contrast with what was then being produced in the USA. Myrdal's 'Prologue' ('The Beam in Our Eyes') is a systematic review of the sources of bias in Western development studies and a strong warning against 'opportunist research' — studies that 'are expected to reach opportune conclusions and to appear in a form that is regarded as advantageous, or at least not disadvantageous, to national interests as these are officially and popularly understood' (p. 12).

austerity, a theoretical power lacking in some of the work of those who attacked it.[17]

Practically, the modernization theorists envisaged modern values being diffused through education and technology transfer to the 'élites' of the periphery. Some attention was paid to this idea in aid policies, especially through technical assistance and scholarship programmes, but on the whole its influence on policy was minor. The modernization school had a bigger impact on academic research, although this owed more to the important topics they opened up by their well-funded fieldwork — topics such as political parties, social movements and the dynamics of social change, whose study had not been encouraged by the former colonial authorities — than to their methodology. And, although the influence of Max Weber on their work was transmitted in the schematized form of Talcott Parsons's 'pattern variables', it had some valuable consequences; for example, some modernization research took seriously the persistence of precapitalist social relations and their cultural practices, issues that were largely neglected by the modernization school's critics in the 1970s.[18]

But modernization theory suffered from defects closely connected with its leading exponents' place in the scheme of things. As Irene Gendzier has pointed out, they were mostly closely connected to the American state and accepted its purposes, including its intense preoccupation with combating communism.[19] Some modernization theorists were serious cold warriors — Gabriel Almond, Edward Shils, Lucien Pye and Samuel Huntington, for example — others merely accepted the Cold War and were content to see themselves as the 'liberal' wing of American development studies, believing that modernization would in any case bring democracy as well as economic growth.[20] Very few at that time publicly questioned the identification

[17] David E. Apter, writing with Carl G. Rosberg, in 'Changing African Perspectives', in Apter and Rosberg (eds), *Political Development and the New Reality in Sub-Saharan Africa* (Charlottesville: University Press of Virginia, 1994), pp. 22–23. The claim to superior clarity and theoretical power that is made here for modernization theory is, to say the least, debatable. The authors do acknowledge, however, that the modernization 'paradigm' had 'some fundamental flaws', including the assumption that each country would progress unproblematically towards modernity, and that the role of external forces would be benign.

[18] In this respect a writer whose work deserved more respect than it got from the left in these years was Fred Riggs, especially his initial study, *Administration in Developing Countries: The Theory of Prismatic Society* (Boston: Mifflin, 1964).

[19] Irene Gendzier, *Managing Political Change: Social Scientists and the Third World* (Boulder: Westview, 1985). This is a central text for any serious understanding of the genesis of American development theory in US foreign policy and anticommunism. The leading exponents of the political development sub-field of modernization theory listed by Gendzier are Gabriel Almond, David Apter, Leonard Binder, James S. Coleman, Karl Deutsch, S.N. Eisenstadt, Samuel P. Huntington, Joseph LaPalombara, Harold D. Lasswell, Daniel Lerner, Max Millikan, Lucien Pye, Edward A. Shils, Sidney Verba and Myron Weiner. American critics of their views were few and far between.

[20] The later illiberal form of modernization theory exemplified by S.P. Huntington was first anatomized by Donal Cruise O'Brien and is further discussed in Chapter 2 below. Gabriel Almond, in his contribution to Wiener and Huntington's *Understanding Political Development*

of modernization studies with the aims of US foreign policy. In the 1950s and early 1960s the Central Intelligence Agency (CIA) regularly 'debriefed' US scholars returning from Third World fieldwork and the State Department frequently sought their advice. This situation also led to a 'symptomatic silence' about the social character of development, a silence cloaked, perhaps, by the doctrine of 'value-freedom'. It was implicit that the development under discussion was not socialist, but its capitalist character was not acknowledged either; it was just 'development', and was certainly not seen as prone to generate class formation and conflict, or as inherently uneven or crisis-ridden.

The shortcomings of modernization theory were first attacked where they were most plainly apparent — in Latin America, which had enjoyed formal independence for more than a century, but had still to enjoy the fruits that according to modernization theory ought long since to have flowed from it. Or, rather, they were attacked from within Latin America by the German-American Andre Gunder Frank, arriving in Chile from the USA in 1962, using the concepts of dependency and underdevelopment. Dependency theory was not invented by Frank (although he played a leading part in its formulation in Latin America in the 1960s); it was implicit in a long tradition of Latin American historiography (which Frank cited extensively) and also in the structuralist analysis developed at the Economic Commission for Latin America under Raoul Prebisch in the 1950s. But Frank was uniquely qualified and motivated to deploy the concepts of dependency against modernization theory, the politics of which he had left the USA to escape.[21] Even before the naïve optimism of much early modernization theory had been exposed by the end of the post-war boom and the deepening US involvement in Vietnam and other anticommunist ventures, Frank's polemical assaults, coinciding with the student revolt of the 1960s, had effectively demolished its pretensions to scientificity.[22]

The early 1970s thus became — briefly — an era of dependency theory. Or, to be more accurate, in intellectual circles, especially among students in Europe and in Third World countries, dependency theory held the initiative; and eventually even the international 'development community' felt obliged to accommodate some of its perspectives: for instance, the International

(op. cit.), argues that most 'modernization' theorists were pessimistic about the prospects for democracy, citing Weber, Gershenkron, Shils, LaPalombara and Wiener. However, there are plenty of counter-examples, including David Apter, James Coleman, Karl Deutsch and Carl Rosberg.

[21] For Frank's intellectual itinerary see his 'The Underdevelopment of Development', special issue of *Scandinavian Journal of Development Alternatives* 10/3, September 1991.

[22] Especially 'Sociology of Development and Underdevelopment of Sociology', first published in 1967, and the other articles collected in *Latin America: Underdevelopment or Revolution?* (New York: Monthly Review Press, 1969), reprinted six times. It is interesting, as I noted in relation to Huntington's work (see Chapter 3 below), how little the US modernization school establishment felt it necessary to respond to their critics, in spite of the effectiveness of the critique in sidelining modernization thinking outside the USA.

Labour Office's 1972 call for 'redistribution with growth' and the World Bank's adoption in 1973 of the principle of meeting 'basic needs' were both influenced by the (unacknowledged) impact of dependency thinking.[23]

Dependency theory inverted many of the assumptions of modernization theory. It saw metropolitan policy as maleficent, not beneficent; inflows of foreign investment were seen as giving rise to much greater interest and profit outflows; 'modernizing élites' were really compradores, or lumpen-bourgeoisies, serving their own and foreign interests, not those of the people; world trade perpetuated structures of underdevelopment, rather than acting as a solvent of them. Capitalist development ('development' now had a label, at least for 'left' dependency theorists) offered nothing to the periphery; the solution lay in reducing links to the metropoles and bringing about 'autocentric' national economic growth.

There were wide differences within dependency theory, of course. 'Bourgeois' dependency theorists in Latin America, such as Osvaldo Sunkel or Celso Furtado, could perhaps be considered (however unfairly, in terms of their actual party-political sympathies) 'organic intellectuals' of their own national capitalist class, chafing at its subordination to the interests of foreign companies and the influence of the US state in domestic politics. Radical or 'left' dependency theorists such as T. Dos Santos and Rui Mauro Marini, on the other hand, aligned themselves openly with the Latin American labour movements, or with radical parties such as the MIR in Chile, and paid a high price by losing their jobs, being forced into exile or even being murdered during the era of the military dictatorships.[24] But both

[23] The prime movers in the former initiative were Hans (now Sir Hans) Singer and Richard Jolly, who led the team responsible for *Employment, Incomes and Equality: A Strategy for Increasing Productive Employment in Kenya* (Geneva: ILO, 1972). Singer and Jolly were both at the Institute of Development Studies (IDS) at the University of Sussex, which bridged (sometimes uneasily) the worlds of academia and practical policy-making in the Third World; at various times the IDS was home to a number of dependency theorists including Manfred Bienefeld, Raphael Kaplinsky, Henry Bernstein and myself. The 'basic needs' initiative came from within the staff of the World Bank, and especially Hollis Chenery; it was officially propounded in 1973 in a speech to the Board of Governors of the Bank meeting in Nairobi by the then president of the Bank, Robert McNamara. Of course, it would be absurd to attribute too much influence to dependency theory as such. It was rather that the growing realization that the hopes of rapid and widespread development were going to be disappointed, and sympathy for the masses in the ex-colonies and fear of their reactions, had been better anticipated by the discourse of dependency, which by this time had become influential everywhere in the Third World.

[24] Some dependency theorists in Latin America who were persecuted under the military dictatorships were supported by foundations and universities in the USA, producing a particular kind of accommodation between Latin American neo-Marxism and American liberalism. The Brazilian *dependencista* Fernando Henrique Cardoso is only a leading example of this; he was very early on considered the 'right' sort of Marxist by leading US academics concerned with the challenge of Marxism. His book with Enzo Faletto, *Dependency and Development in Latin America* (Berkeley: University of California Press, 1979), was rightly seen as a model of the creative use of a neo-Marxist conceptual framework for analysing the diverse 'concrete situations of dependency', which did not commit the authors to a particular political line and managed to combine inspiring flourishes with a generally dispassionate

tendencies had one thing in common, which separated them from all the other schools of development theory so far discussed: their intellectual debt to Marx and Hegel. Their critique of official development thinking rested fundamentally on a pulling away from the short-term, ahistorical and uncritical perspectives of Western-produced, state-orientated development discourse, towards the perspective of a 'universal history'.

But they themselves also believed that the countries of the 'periphery' could somehow, through better theory and different political leadership, jump over the barriers placed in their way by history, and this gave rise to some key ambiguities in their thought: above all, their tendency to assume the availability of some unspecified alternative development path, more equitable and less painful, which was not — in the absence of stronger and more mobilized social forces at the periphery, and more sympathetic support from abroad — really available.[25] This problem persisted, even when Frank's early version of dependency theory, according to which development was always systematically blocked at the periphery, had been generally abandoned in favour of the idea that, while it was always necessarily difficult, dependent on external forces and 'distorted' (Cardoso's famous 'associated dependent development'), development might none the less sometimes be possible.[26]

In sub-Saharan Africa, dependency theory was broadly accepted by many foreign Africanists and many, perhaps most, African social scientists, not to mention educated people in general, and especially the youth;[27] but there

tone. For many Americans, as for John Toye (in *Dilemmas of Development: Reflections on the Counter-revolution in Development Economics* (Oxford: Blackwell, second edition 1993), pp. 128–30) and Gabriel Palma ('Dependency: A Formal Theory of Underdevelopment or a Methodology for the Analysis of Concrete Situations of Underdevelopment?', *World Development* 6, 1978, pp. 881–924), this was the kind of Marxism to back. In 1994, Cardoso, although himself a founder of the Brazilian Social Democratic Party, would stand successfully for the Presidency of Brazil with the support of business against the left's Lula da Silva.

[25] This was the most telling theoretical criticism in Bill Warren's influential article, 'Imperialism and Capitalist Industrialisation', *New Left Review* 81, 1973, pp.3–44. The definitive discussion of this point is in Gavin Kitching, *Development and Underdevelopment in Historical Perspective* (London: Methuen, 1982).

[26] F.H. Cardoso, 'Dependency and Development in Latin America', *New Left Review* 74, 1972, pp. 83–95. One of the clearest, and perhaps one of the most influential, assessments of dependency theory was Gabriel Palma's 'Dependency: A Formal Theory of Development or a Methodology for the Analysis of Concrete Situations of Underdevelopment?', op. cit.

[27] Bjorn Beckman's observation on dependency (or 'underdevelopment') theory in relation to Africa at the beginning of the 1980s was no doubt true, however: 'Academics may have contributed in articulating it but the tremendous diffusion of its perspective can only be understood as a response to specific historical experiences and the development of social forces at the world level, including the realities of colonialism and neo-colonialism, the rise of socialist countries and armed liberation struggles. It is not a specific political line with a uniform theoretical basis. It is a position held by millions of anti-imperialist militants most of whom may never have heard of or read the works of Andre Gunder Frank or Samir Amin' (Bjorn Beckman, 'Imperialism and the "National Bourgeoisie"', *Review of African Political Economy* 22, 1981, p. 10). It would be interesting to try to assess how far Africans' experience since 1981, as interpreted to them by the leadership of the age of structural adjustment, has destroyed this widespread understanding. Are there still millions of anti-imperialist militants?

was a further problem, that outside the Republic of South Africa the level of development in few countries had yet produced either a local 'national' capitalist class or a local labour movement (or indeed any other modern social movement) that had the capacity to lead national development along any alternative development path, even if such a path could be plausibly specified.[28] As a moral critique of existing policy in Africa, dependency theory played a significant role.[29] But, except in Nyerere's Tanzania, dependency thinking was not adopted as an explicit basis for policy, and the problems of Tanzanian socialism had many sources besides the inherent shortcomings of dependency theory.

In any case, it was not shortcomings revealed in practice that led to the most significant critiques of dependency theory.[30] Critics from the right generally failed to make the effort needed to understand the Marxian *problématique* from which many of dependency theory's key ideas were drawn, so as to be able to make effective attacks on it (this was particularly evident in the attempts to use cross-national statistical data to prove, for example, that periphery country growth rates were not inversely related to trade links with the countries of the 'core').[31] The most damaging

[28] This judgement may seem to ignore the potential of Africa's peasantries, especially those who were forced into prolonged armed liberation struggles; and it may well be that history will judge the defeats inflicted on these peasantries' aspirations by global forces in the 1980s as one of the most crucial contributions to the continent's end-of-the-century morass. For the time being, however, the original potential of Africa's peasant-based socialist projects is very hard to assess with any confidence.

[29] Ngugi wa Thiong'o's books such as *Petals of Blood* (London: Heinemann Educational Books, 1971) and *Devil on the Cross* (London: Heinemann, 1982) perhaps represent the deepest popular impact of dependency thinking, i.e. as a populist moral discourse.

[30] The statist development strategies that were adopted by socialist governments in Guinée, Ghana, Mozambique and elsewhere were mostly informed as much by neo-Stalinist ideas about development, contributed by East European aid personnel, as by dependency theory, although in each case the mix was complex and the outcomes less predictable than they are often made to seem in retrospect. The Tanzanian experience did eventually become a focus for critics of dependency theory but only long after the debate had been initiated on more general grounds, and some time after the 'Kenya debate'. See especially the incisive contribution of Gavin Kitching, *Development and Underdevelopment in Comparative Perspective: Populism, Nationalism and Industrialisation* (London: Methuen, 1982), especially Chapter 5; and also Suzanne Mueller, 'Retarded Capitalism in Tanzania', *Socialist Register* 1980, pp. 203–26, and 'The Historical Origins of Tanzania's Ruling Class', *Canadian Journal of African Studies* 15/3, 1981, pp. 459–97.

[31] The conclusions of Jorge Larrain on this point cannot be improved upon: 'There is little doubt that if one examines the theory of dependency...as a testable theory with precisely defined variables and concepts whose characteristics are exclusive and apply only to dependent countries and one requiring measurable empirical evidence which substantiates the hypotheses, then the theory of dependency does not seem to pass the test...But one wonders whether this attempt to judge the theory of dependence in accordance with such a formal pattern is really worthwhile. Such an attempt is bound to abstract from and miss what had been the essence of the theory in its origins and in the intention of its best representatives: the historical analysis of Latin American processes "as the result of struggles between classes and groups that define their interests and values in the process of expansion of a mode of

criticism came rather from the 'classical Marxist' left.[32]

These critics were, ironically enough, probably the nearest thing we now have to 'traditional intellectuals', in Gramsci's sense of the term (i.e. a category of intellectuals not linked to either of the main contesting classes). Their mostly marginal positions, even in European academic life (not to mention the USA, where even in the mid-1970s few campuses would give them a home), tended to detach them from metropolitan social forces and interests concerned with the Third World, while the fact that they were typically not engaged as consultants meant that many of them, after an initial period of teaching at some Third World university, also tended to become separated from social forces and interests in Third World countries as well. Writers like Geoffrey Kay, Giovanni Arrighi, Arghiri Emmanuel, Michael Cowen and Bill Warren often seemed to display the attachment of the political exile (in their case, exile from the academic and policy-making mainstream) to theory as such. While this had its disadvantages, it did enable them to make a trenchant critique of the eclecticism, populism and practical ambiguity of dependency theory: now for the first time 'development theory' of the post-war variety was squarely confronted from the perspective of the historical tradition of development theory derived from Hegel and Marx.

To take a well-known early example, in 1975 Geoffrey Kay showed that *dependencistas* who accused the metropoles of 'exploiting' the periphery rarely had a very clear idea of what they meant by this. Marx, in contrast, had formulated a clear concept according to which capitalist exploitation referred to the appropriation by the owners of capital of the surplus value produced by workers; and the more surplus value they produced, the greater the rate of exploitation. The most exploited workers, on this analysis, are those that produce most surplus value, and they are the ones who work with most capital, such as the highly paid workers of the US automobile industry. Therefore the real problem of the Third World, Kay maintained, was not that it was superexploited, but that it was not exploited enough.[33]

Like many of the theses advanced by the classical Marxists in their contributions to development theory, however, this was a 'Menshevik' kind of

production" and which "in the struggle for control or for the reformulation of the existing order... are making a given structure of domination historically viable or are transforming it"' (Jorge Larrain, citing Cardoso, in Larrain, *Theories of Development: Capitalism, Colonialism and Dependency* [Oxford: Polity Press, 1989], pp. 178–79).

[32] I use the term 'classical' here, rather than 'orthodox', which has pejorative overtones, to designate those Marxists who saw themselves as applying and perhaps extending but not revising or modifying Marxist theory.

[33] 'The radical critics of orthodox development theory were so keen to prove the ideological point that underdevelopment was the product of capitalist exploitation, that they let the crucial issue pass them by: capital created underdevelopment not because it exploited the underdeveloped world, but because it did not exploit it enough' (Geoffrey Kay, *Development and Underdevelopment: A Marxist Analysis* [London: Macmillan, 1975], p. x). The same point was made by Arghiri Emmanuel in 'Myths of Development versus Myths of Underdevelopment', *New Left Review* 85, 1974, pp. 61–82.

argument, in the sense that it seemed to imply that capitalist development of the periphery was a necessary prelude to socialism: not a political stance that appealed to many people on the left, inside or outside the Third World, who in any case did not believe it would happen. This was a central difficulty in the classical Marxist position and it was compounded, rather than resolved, by the position eventually adopted by Bill Warren, who argued in his book *Imperialism: Pioneer of Capitalism*, published posthumously in 1980, that capitalism was in fact developing rapidly throughout the Third World, and accomplishing its 'civilizing' mission there into the bargain.[34] This hardly corresponded to what many people in the Third World (and especially in Africa) seemed to be experiencing, and certainly not to the feelings most people had about capitalist development, even in places (such as, for example, Mexico or India) where it might seem to be occurring. Perhaps Warren, had he lived to revise his book for publication, would have avoided giving the impression that capitalist development could be expected to develop the entire world, and soon. As it was, his text bent the stick much too far in the opposite direction from the early dependency line.

Marxist development theorists were also frequently attacked for being 'Eurocentric', especially for applying to backward societies categories like that of 'the working class', which did not apply there, and neglecting phenomena like ethnicity, which did. On the whole, this was a canard. The real issue was how far capitalist development was forming classes, and how far this cut across ethnic and other precapitalist solidarities. In practice, the empirical studies conducted by Marxist researchers were no more Eurocentric than those of their critics.

No, the real problem of the Marxists' contribution to development theory was not so much that their analysis was wrong; in many ways they appear in retrospect to have maintained a rather objective stance, relative to the various other schools, helped by the broad historical perspective and understanding of capitalist dynamics that they drew from Marx. Their crucial problem was rather that there were too few people in the Third World — and virtually none in tropical Africa — for whom the political and moral standpoint of their analysis (i.e. that people should struggle against capitalist development, while not expecting to transcend it until it had first been accomplished) made sense.[35] Their perspective was, to say the least, very

[34] Warren seems to have seen his position as Leninist, i.e. Bolshevik, rather than Menshevik, to judge from his strong endorsement of Lenin's position in the debate with the populists and the Mensheviks in Russia (Bill Warren, *Imperialism: Pioneer of Capitalism* (London: Verso, 1980), pp. 30–39. But his unfinished text gave the impression that he held the wholly implausible view that Russia's development trajectory down to 1917 was bound to be repeated throughout the Third World in the 1980s and 1990s.

[35] P.W. Preston comes close to making this point among many other penetrating observations in his otherwise maddeningly ill-digested book, *Theories of Development* (London: Routledge and Kegan Paul, 1982), where he says (p. 177) that one criterion of a good theory (or ideology, in his usage) must be: 'Does the stance in question specify an agent of theory-execution whereby the effort can latch on to the real world?'

long-term, and offered no plausible line of immediate political action to improve matters.[36] The fact that 'mainstream' development theory had consistently failed to produce results did not make the Marxist view any better in this respect.[37]

A more plausible political position was, of course, that of the neo-liberals, who did not believe that capitalism would give way to socialism and were only interested in accelerating its advance in the Third World.[38] They believed that what was blocking or retarding this was none of the things highlighted by all the theories so far discussed, but rather the whole idea of bringing about development through state intervention in the economy in the first place. This was the standpoint of P.T. (later Lord) Bauer, Deepak Lal, Bela Balassa, Ian Little and others, who represented in development theory the

[36] Gavin Kitching made a characteristically honest attempt to face this issue when he said: 'I believe that a socialist alternative to capitalism in Africa *can* be constructed by a sophisticated socialist movement created among the working classes of Africa as they *slowly* expand through time...I believe, however, that the formation of a sophisticated socialist working class in Africa will take a long time and that a prolonged period of struggle against a developing capitalism there is one of the important prerequisites of its creation. To that extent I am "happy" to see continued capitalist development in Black Africa (though this may or may not occur)' ('Politics. Method and Evidence in the "Kenya Debate"', in Henry Bernstein and Bonnie K. Campbell (eds), *Contradictions of Accumulation in Africa* (Beverly Hills: Sage, 1985), pp. 147–48. African readers of this might well wonder how socialism would come about if continued capitalist development, against which they should be struggling, did not in fact occur to the point where a 'sophisticated working class' existed.

[37] For an engaging discussion of the spurious and self-interested demand often made by some members of the 'development community', that critical theorists of development should come up with an answer to the question of 'what is to be done?', see James Ferguson, *The Anti-Politics Machine: 'Development,' Depoliticization, and Bureaucratic Power in Lesotho* (Cambridge: Cambridge University Press, 1990), pp. 279–88. Members of the 'development industry', he remarks, 'seek only the kind of advice they can take. One "developer" asked my advice on what his country could do "to help these people". When I suggested that his government might contemplate sanctions against apartheid, he replied, with predictable irritation, "No, no! I mean *development*!" The only "advice" that is in question here is advice about how to "do development" better' (p. 284).

[38] The convergence of Marxist and neo-liberal thinking in terms of their analysis of global capitalism should not have surprised anyone who appreciated their shared intellectual roots in the thought of Smith and Ricardo, as Dudley Seers pointed out in *The Congruence of Marxism and Other Neo-classical Doctrines*, IDS Discussion Paper No. 13 (University of Sussex: Institute of Development Studies, 1978). While the outrage with which many 'progressives' greeted the publication of Warren's book was due in part to some deliberate provocation on his part, it sometimes also seemed to be a symptom of some unease with regard to the critics' own conception of the route to Third World socialism, which was not necessarily very clear or plausible; see among others the polemical attack by Alain Lipietz in *New Left Review* 132, 1982, pp. 48–58. Conversely, a similar observation might be made about John Toye's attempt to blame the 'important intellectual failures' of Marxist development theory for facilitating the 'counter-revolution' of the neo-liberals (*Dilemmas of Development*, op. cit., pp. 132–33). To the extent that intellectual factors were at all significant in the neo-liberal triumph, this seems a curious exaggeration of the influence of Marxism. Given the near-monopoly of positions of influence in the 'development community' possessed by non-Marxists like Toye himself, one might think that a greater responsibility to avert the counter-revolution rested with them.

neo-liberal revolution that was taking place in the metropoles at the end of the 1970s, and who offered an intellectual justification for a new wave of market-orientated intervention by the World Bank and the International Monetary Fund (IMF). The older representatives of this current belonged to a small group of economists who opposed the post-war social-democratic consensus and who were, as a result, almost as exiled from the mainstream as the Marxists (although sustained over the years, as Radhika Desai has shown, by membership in a sect-like network organized by the Mont Pelerin Society, the American Enterprise Institute and the Institute of Economic Affairs).[39] They argued that development was blocked by inflated public sectors, distorting economic controls and overemphasis on capital formation.[40] Governments were part of the problem, not part of the solution; they were inefficient and often corrupt and hence parasitic, not stimulators of growth. The solution was to privatize the public sector, reduce the scale and scope of government spending and give up all policies, from exchange rate controls to subsidies and redistributive taxation, that altered any prices that would otherwise be set by the impersonal forces of the market.

As John Toye pointed out, the neo-liberals' success in relation to the Third World owed a good deal to the fact that they were ready to say openly what others in the 'development community' knew perfectly well but had (unlike the dependency theorists) been unwilling to say, out of an anxiety not to jeopardize relations between Third World governments and the development agencies for which they worked: namely, that these governments were never exclusively concerned to promote the development goals they were ostensibly committed to, and quite often were not committed to them at all.[41]

There was also a strong core of justification for their criticisms of the public sector and of government practices in most Third World countries. But, as Toye has also shown, this common-sense criticism did not add up

[39] Radhika Desai, 'Second Hand Dealers in Ideas: Think Tanks and Thatcherite Hegemony', *New Left Review* 203, 1994, pp. 27–64.

[40] Lord Bauer has outlined his early thinking in 'Remembrance of Studies Past: Retracing First Steps', in Gerald M. Meier and Dudley Seers (eds), *Pioneers in Development* (New York: World Bank and Oxford University Press, 1984), pp. 27–43. The general position of the neo-liberals is summarized and criticized by John Toye in *Dilemmas of Development*, op. cit., Chapters 3 and 4. See also Christopher Colclough, 'Structuralism versus Neo-liberalism: An Introduction', in C. Colclough and James Manor (eds), *States or Markets? Neo-liberalism and the Development Policy Debate* (Oxford: Clarendon Press, 1992), pp. 1–25.

[41] 'As professional economists seeking to promote reforms, they assumed the existence of certain institutions and attitudes, as it were trying to coax them into life while aware that they were often not in fact there... Many felt morally uncomfortable in their inability to explore openly the reasons for their professional frustration, but most of these loyally respected the diplomatic imperative' (*Dilemmas of Development*, op. cit., p. 143). Dependency theorists, in contrast, had from the first emphasized the tendency of Third World élites in power to serve their own and foreigners' interests at the expense of their people's collective interests. Unlike the neo-liberals, on the other hand, they did not make the mistake of assuming that no Third World governments were ever genuinely concerned for the public interest.

to a theoretical justification for the neo-liberals' claims about the benefits that would flow from an unrestricted market. These claims were very poorly supported with evidence, and were often prima facie implausible; they sprang rather from a deep ideological hostility to government in general, and especially to the legitimacy which the doctrine of state intervention gave to socialists or even social-democrats in office.

In any case it was not the shortcomings of the principal existing schools of development theory, serious as they were, that made possible the ascendancy of neo-liberalism (whose shortcomings were quickly revealed as no less serious). What made possible the triumph of neo-liberalism in mainstream development thinking was material, not ideal: the radical transformation in both the structure and the management of the world economy that had begun in the 1960s, and which finally seemed to offer the possibility of creating for the first time in history a truly unified global capitalist economy — and one regulated, if at all, only by institutions reflecting the interests of transnational capital. Neo-liberalism articulated the goals and beliefs of the dominant forces that stood to benefit from this process, and pushed it forward. Social-democratic parties and labour movements tried to resist it, but the 'new right' succeeded in neutralizing this resistance and initiating its own market-orientated project in one industrial country after another.[42] The 'development community', which was either part of the state apparatuses of these countries or depended critically on them for funding, was bound to come into line.[43]

But, although the 'develoment community' was loath to acknowledge it, the new global economic regime thoroughly undermined the foundations of development theory as it had hitherto been conceived.

The real world of development

The world in which Keynesian policy-making — and its offshoots, development economics and development theory — made sense had changed

[42] The process was led by the UK and the USA but gradually accepted by the other OECD countries, the failure of France to resist the pressure in 1983 being a key turning-point; for a succinct review of this process see Eric Helleiner, 'From Bretton Woods to Global Finance: A World Turned Upside Down', in Richard Stubbs and Geoffrey R.D. Underhill (eds), *Political Economy and the Changing Global Order* (Toronto: McClelland and Stewart, 1994), pp. 163–75.

[43] The dependency of international agencies such as the World Bank and the IMF on the policy preferences of their major participating states is obvious, but the pressures extended to the non-government organizations (NGOs) in the development field as well. In Canada at the end of the 1980s, for example, NGOs involved in development depended on Canadian International Development Agency (CIDA) funding for about 40% of their revenues. As CIDA adopted the federal government's increasingly strong neo-liberal line it switched its support away from NGOs that resisted this to more uncritical funding channels, such as Canadian universities; see Tim Brodhead and Cranford Pratt, 'From Friction to a Fresh Start: CIDA, the NGOs and Foreign Policy', *Briefing* (Ottawa: The North–South Institute, 1944 – B37).

fundamentally.[44] It is true that in some respects the world economy at the end of the 1980s was less integrated than it had been at the beginning of the century, and there were significant tendencies towards protectionism, offsetting those towards a single worldwide market.[45] But, relative to the situation that existed between 1945 and the late 1960s, the changes were fundamental.

World trade as a share of world output had returned to the general level of 1913 (i.e. up from 7% of total gross domestic product (GDP) in 1945 to 15% in 1988); foreign direct investment had risen to account for significant shares of total investment (5-10% of capital stock) in most major economies, and about a third of all trade between countries had come to consist of the movement of goods between different national branches of one or another multinational company. The mutual dependence of national economies implied by these facts was significant (obviously, it is not necessary for half of a country's capital assets to be foreign owned for decisions taken in foreign countries to have a major impact on its fortunes); but even more striking was the internationalization of capital flows. Instead of merely financing world trade, by the end of the 1980s banks and non-bank financial institutions were dealing in currency exchanges, currency and commodity futures and so-called 'derivatives' of all sorts on a scale that not only dwarfed the conventional transactions needed for trade and investment, but made it impossible for the governments of even large economies to influence the value of their currency by intervening in the currency markets.[46]

[44] Andrew Glyn and Bob Sutcliffe, 'Global But Leaderless? The New Capitalist Order', *Socialist Register 1992* (London: Merlin Press, 1992), pp. 79–81. See also Harry Magdoff, 'Globalization: To What End?', ibid., pp. 44–75; David Gordon, 'The Global Economy: New Edifice or Crumbling Foundations?', *New Left Review* 168, 1988, pp. 24–64; Susan Strange, *Casino Capitalism* (Oxford: Blackwell, 1986, pp. 1–102); James Crotty, 'The Limits of Keynesian Macroeconomic Policy in the Age of the Global Marketplace', in Arthur MacEwan and William K. Tabb (eds), *Instability and Change in the World Economy* (New York: Monthly Review Press, 1989), pp. 82–100, and Gerald Epstein, 'Financial Instability and the Structure of the International Monetary System', ibid., pp. 101–20.

[45] Glyn and Sutcliffe (in 'Global But Leaderless?', *Socialist Register 1992*, op. cit., pp. 76–95) and other commentators are careful to make this point, relying especially on R. Zevin, 'Are World Financial Markets More Open? If So, Why and With What Effects?' in T. Banuri and J. Schor (eds), *Financial Openness* (Oxford: Oxford University Press, 1992). The World Bank reported that 'In 1986, more than 20% of imports from the developing countries were covered by "hard-core" [non-tariff protection] measures alone' (*World Development Report* (Oxford: World Bank, 1991), p. 8); the Bank of International Settlements estimated that 50% of world merchandise trade is 'managed' in ways that evade GATT rules (cited in Harry Magdoff, 'Globalisation: To What End?' op. cit., p. 50).

[46] A classic example was provided by the collapse of the exchange rate mechanism (ERM) of the then European Community in September 1992, when the British government tried to maintain the pound sterling in the ERM by the traditional expedient of using the country's reserves of foreign currencies to buy pounds and so create a demand for it, to offset the selling by speculators counting on forcing a devaluation. The government spent several billion pounds in less than two days in this effort: the volume of currencies traded on the world's currency exchanges on an average day at that time was three trillion dollars.

But in the meantime control of capital movements had in any case also been abandoned as a deliberate policy decision, promoted, above all, by the USA. As competition with other industrial countries intensified, the USA borrowed abroad and became the world's largest debtor nation. As such it was obliged to give up its commitment to meet its dollar obligations in gold, which it did in 1971;[47] and eventually, in 1973, the Bretton Woods system of fixed exchange rates was abandoned altogether. This opened up new opportunities for international currency speculation and led to a new period of extreme instability in currency values and commodity prices, including the oil price increases of the 1970s and 1980s; these in turn led to vast new dollar balances being accumulated by the oil-exporting countries, and correspondingly vast expansions of borrowing, which drove up the total of international debt to previously unheard-of levels.

The abandonment of the post-war international trading regime was followed in 1979-80 by the abandonment of Keynesian economic policy in the Organization for Economic Co-operation and Development (OECD) countries, led by the UK and the USA. Deregulation in the USA and deregulation and privatization in the UK were accompanied by high interest rates. The governments of the other European industrial countries followed suit, either willingly or (in the case of France) because keeping interest rates significantly below those of other countries led to capital outflows that could no longer be prevented — 'Keynesianism in one country' was no longer practicable. Capital exports were formally deregulated in the UK in 1979 and *de facto* everywhere else by the mid-1980s. Then, at the end of 1993, the conclusion of the 'Uruguay Round' of the General Agreement on Tariffs and Trade (GATT) negotiations inaugurated a further extension of global free trade, including the formerly sacrosanct agricultural sector, while further reductions in the regulatory powers of most industrial country governments were imposed by the European Union's Single Market and Maastricht Treaties and the North American Free Trade Agreement.

These changes did not succeed in restoring growth rates to the levels achieved after the Second World War. From the late 1960s the average rate of growth of the OECD countries fell from the post-war level of 3-4% to around 2%. The developing countries inevitably followed suit, except that there was now a growing polarization among them. Besides the four East Asian newly industrialized countries (NICs) (which accounted for half of the entire Third World's exports of manufactures), in the 1980s China and, to a lesser extent, India began to grow faster, while the other developing countries slowed down — in the recession

[47] As Gerald Epstein points out, the USA had already used its political and military leverage to avoid having to honour this commitment well before it was obliged to abandon it formally (see Epstein, 'Financial Instability and the Structure of the International Monetary System', in MacEwan and Tabb (eds), *Instability and Change in the World Economy*, op.cit., pp. 101–20.

that began in the late 1970s their average growth rate declined and in 1983 even became negative.[48] Lower growth rates in the OECD countries and intensified competition also adversely affected the Third World countries' terms of trade and interest rates: for the 1980s, 'in Sub-Saharan Africa and Latin America the combined effects [of declining terms of trade and interest rate increases] were estimated to average more than 10 percent of GDP'.[49]

Most Third World countries, then, found themselves more vulnerable than at any time since they were first colonized. Their economies were least well placed to prosper in the new 'global' market place. Primary commodity exports, other than oil, became steadily less significant as manufacturing became less commodity-intensive, and the overall share of the Third World in world trade fell dramatically.[50] Faced with stagnating economies, and with per capita incomes declining from levels at which many people could barely survive already, they responded by increased borrowing abroad until servicing the debt led to balance of payments difficulties so acute that they were forced to turn to the IMF. As a condition of further support the IMF and the World Bank then forced them to cut back government intervention in their economies, leaving these instead to be revived by the freer play of 'market forces'. This did not, of course, produce the anticipated results. Per capita incomes fell still further in all the affected countries (in sub-Saharan Africa, by over a quarter), while the debt-service burden (the proportion of export earnings spent on capital repayments and interest) of the 'low-income countries' (excluding China and India) rose from 11.8% in 1980 to 24.5% in 1992.[51] The overall effects are well summed up by Glyn and Sutcliffe:

> The share of Africa, Asia and Latin America in world trade is now substantially lower than before 1913. This reflects a major decline in the relative importance of tropical raw materials in world trade... This long-term structural reason for the decline...has been joined in the period since 1973 by a major short-term crisis in many poorer countries... Latin America's share of world exports fell from 12.4 percent in 1950 to 5.5 percent in 1980 and only 3.9 percent in 1990; Asia's was 13.1 percent in 1950, rose to 17.8 percent in 1980 and fell back again

[48] According to the Bank's *World Development Report* for 1994, the low income countries as a whole, excluding China and India, experienced an average annual per capita growth rate of 1.2% between 1980 and 1992, while the middle-income countries had an average annual decline of 0.1%. By regions, per capita incomes in sub-Saharan Africa declined by 0.8% a year on average in the same period, in the Middle East and North Africa they declined by 2.3% and in Latin America and the Caribbean by 0.2%.

[49] *World Development Report* 1991, p. 18.

[50] The four South-East Asian NICs (Korea, Taiwan, Hong Kong and Singapore) account for almost half of all Third World exports of manufactures; the share of the rest of the Third World has declined over the past two decades.

[51] Manfred Bienefeld, *Rescuing the Dream of Development in the Nineties*, Silver Jubilee Paper 10 (University of Sussex: Institute of Development Studies, 1991, p. 13): a terse and powerful critique of the course and effects of structural adjustment policies.

> to 14 percent in 1990. And Africa's share all but disappeared: in 1950 it was 5.2 percent, in 1980 4.7 percent and in 1990 only 1.9 percent. In 1955 exports from Africa were three times those of Japan; by 1986 the exports of Japan were four times the exports of Africa. The picture for international investment is rather similar. Between 1950 and 1980 the share of all foreign investment going to the Third World held roughly constant at about 25 percent. But after 1984 the share fell sharply to well under 20 percent... [and] is very unequally distributed. It goes in significant quantities to only a few resource-rich countries and the newly industrialising countries (including China) while the so-called 'least developed countries' are increasingly excluded. In the second half of the 1980s this group received only 0.1 percent of all foreign investment... Once again Africa and most of Latin America and some Asian countries are failing to participate in the growing globalisation of the rest of the world... they are increasingly marginalised within the system of which they form a part.[52]

The story of the world economy under liberalization can, of course, be given a rosy gloss, as in the following excerpt from a *Washington Post* editorial:

> The rise of wealth in the late 20th century has been more sustained and more widespread than ever before in history... Economic growth is measured in dollars, but it translates into other and much more important things — better health and longer lives, less harsh physical labor, greater economic security. There are drawbacks, like development's threats to the environment and the dismaying tendency of governments to spend too much of their new wealth on weapons. But it is hardly Pollyannaish to say that the balance remains strongly in favor of the essential human values.[53]

Admittedly, to take this line involves overlooking the implications of many of the data reviewed in the last few paragraphs, not to mention the serious risk of conflicts within or between countries of the former 'First' or 'Second' Worlds as the impact of global competition drives whole districts, regions or even countries into permanent poverty, while others prosper. But for present purposes it does not really matter: even on the most optimistic view this picture leaves little or no room for 'development theory' as it used to be conceived.

The era of national economies and national economic strategies is past — for the time being, at least. With capital free to move where it wishes, no state (and least of all a small poor one) can pursue any economic policy that the owners of capital seriously dislike. Economic planning, welfare systems and fiscal and monetary policies all became subject to control, in effect, by the capital markets, signalled, in the case of Third World countries, by the conditions attached to IMF/World Bank lending —

[52] Glyn and Sutcliffe, 'Global But Leaderless?', op. cit., pp. 90–91.

[53] 'A World Growing Richer', *Washington Post* editorial in the *Manchester Guardian Weekly*, 26 June 1994. The editorial adds: 'The countries that have been left out of this surge fall mainly into two categories: the ones that have been entangled in long wars, and most of those in Africa.' The USA's role in these wars, and in the arms trade, is not mentioned in the editorial.

precisely the situation the Bretton Woods system was designed to prevent.[54] And in the Third World the whole thrust of recent IMF/World Bank policy, imposed through the conditions attached to almost two hundred structural-adjustment lending programmes and reinforced by bilateral lending consortia, has been to reduce still further the power of national governments to act as prime movers of development. Instead of reforming inefficient agencies, structural adjustment policies have tended to emasculate or eliminate them. Parastatals have been privatized, without thereby becoming more effective. It is hardly too much to say that by the end of the 1980s the only development policy that was officially approved was not to have one — to leave it to the market to allocate resources, not the state. In the World Bank's own ingenuous language, 'New ideas stress prices as signals; trade and competition as links to technical progress; and effective government as a scarce resource, to be employed sparingly and only where most needed.'[55]

Individual national governments — especially in the smaller underdeveloped countries, with which development theory used to be above all concerned — thus no longer have the tools at their disposal to manage their domestic economies so as to accelerate growth, foster industrialization and 'catch up', as development theory originally envisaged, and theories premissed on their existence become irrelevant; for most of them Gunder Frank's comment was painfully accurate: 'Now neo-liberalism, post-Keynesianism, and neo-structuralism have...become totally irrelevant and bankrupt for development policy. In the real world, the order of the day has become only economic or debt crisis management.'[56]

Most observers accept that significant parts of the former Third World, including most of sub-Saharan Africa, are more likely to regress than to advance in the new global economy; it is in the nature of an unregulated competitive system that this will happen. Not every country has the capacity to compete in the market; a few will succeed, while others will decline and some will collapse into civil war or anarchy, as Uganda, Somalia, Angola,

[54] See Helleiner, 'From Bretton Woods to Global Finance', in Stubbs and Underhill (eds), op.cit., pp. 164–65.

[55] This revealing formulation comes from the conclusions of the World Bank's review of 'the evolution of approaches to development' in its 1991 *World Development Report* under the heading 'The Way Forward'. As the real consequences of structural adjustment became inescapably obvious towards the end of the 1980s, especially in Africa, official policy veered back towards an emphasis on the importance of the state, but without reconciling this with the continued official emphasis on the superiority of markets over state action — in a word, official policy became contradictory, as any thoughtful reader of the 1991 *World Development Report* can see.

[56] 'The Underdevelopment of Development', op. cit. (note 21), pp. 48–49. The situation is not essentially different in the industrialized countries, as Susan Strange remarks: 'Political leaders and their opponents like to pretend that they are still in control of their national economies...But recent years have shown again and again how the politicians' plans have been upset by changes that they could not have foreseen in the world outside the state', in *Casino Capitalism* (Oxford: Blackwell, 1987), p. 3.

Mozambique, Liberia and Rwanda already have, in Africa, at different times.[57]

A measure of just how profoundly the world has changed may be had by comparing this reality with the following optimistic words written by Sir Arthur Lewis, one of the founding figures of development economics, as late as 1983:

> All LDCs [less developed countries] are menaced since 1973 by the international recession, by rising protectionism in the industrial countries, by the high price of oil, and by the enormous debt this has created. These disasters are beyond their control and call for special measures by the whole international community. But the viability of LDCs in normal times, like the 1950s and the 1960s, is now beyond doubt.[58]

We can now see that the 1950s and 1960s were not 'normal' times but, on the contrary, a special interlude in the history of the worldwide expansion of capitalism in which 'development theory' could be born, but outside which it could not survive.

This is not to say that theorizing development is no longer possible or necessary; we need theoretical maps of our increasingly integrated world. But we can no longer assume, as all the principal varieties of 'development theory' have up to now, who the agents of collective action for change will be, or that means exist for them to accomplish anything. Perhaps states, acting singly or in groups, will rediscover the means, but this too must be part of the task of theory to establish. In the meantime, we must recognize that an era is closed, that development theory must return to its classical roots and that the relation between theory and practice that has been assumed hitherto (i.e. theory in the service of this or that existing or imagined coalition of political forces in control of a state) has been put radically in question.

'Development theory' faced with the end of its *raison d'être*

The authors of the World Bank's annual *World Development Reports* have dealt with the problem by ploughing ahead with an increasingly incoherent

[57] Robert Kaplan's 'The Coming Anarchy', *Atlantic Monthly*, February 1994, pp. 44–76, registered a thoughtful journalist's recognition of the instability now characterizing the global system, although without any apparent understanding of, or even curiosity about, its causes. A more insightful account is Christopher Hitchens, 'African Gothic', *Vanity Fair* November 1994, pp. 92–117. It can hardly be irrelevant that before the outbreak of genocide in Rwanda in 1994, per capita incomes, already among the lowest in the world, had fallen by 50% in one year as a result of a collapse in coffee prices, followed by a drastic rise in food prices imposed by an IMF/World Bank structural adjustment programme.

[58] 'Development Economics in the 1950s', in Meier and Seers (eds), *Pioneers in Development* (op. cit.), p. 137.

discourse of opposites: the state is needed, after all, but not too much, and only when the market doesn't work well; democracy is important but not if it leads to inappropriate demands for redistribution;[59] and so on. Academic development theorists could hardly follow suit, but what could they do instead?

It was not until towards the end of the 1980s that the full significance of the changed environment began to be registered in the theoretical literature, although the drastic reduction in the official goals of development propounded by the World Bank and other agencies over the years had signalled it clearly enough. By the early 1970s the vision of 'catching up' (culminating, in Rostow's 1960 version, in a 'high mass-consumption' society, which implicitly included 'equity' and democracy) had already given way to more modest ambitions: 'redistribution with growth' — i.e. some reduction in inequality, but financed out of growth so that the better off in the developing countries might be less unwilling to agree to it — in a word, fewer illusions about democracy. And by the end of the 1970s redistribution had given way to just trying to meet the 'basic needs' of the poor, who, it seemed, would always be with us after all; the goal of equity had disappeared. Then came structural adjustment; to get growth, underdeveloped societies were to adjust themselves to the procrustean bed allocated to them by the market, and for this purpose even basic needs must be sacrificed.

By then, however, everyone was aware that things had radically changed and that 'development theory' was in deep trouble. Apart from neo-liberalism itself, five main lines of theoretical response can be identified. One has been to see the problem as essentially one of theory itself: there is a theoretical 'impasse', which must be overcome by better concepts and research. A second response might be called 'eclecticism as usual in the development community'. A third consists of further evolutions of dependency theory. A fourth response is to return to the unfinished agenda of Myrdal and the 'neo-institutionalists' of the 1960s — i.e., how to 'add' social and political dimensions on to the analyses of development economics — but this time by analysing these dimensions in terms of rational choice theory. A fifth response, and the last to be considered here, is to renounce any commitment to development, seeking (often in the name of post-structuralism) merely to 'understand' what goes on.

Let us look briefly at these, taking one or two representative thinkers in each case.

[59] For a scathing analysis of the 'development community's' attitude to democracy see Gerald G. Schmitz, 'Democratization and Demystification: Deconstructing "Governance" as Development Paradigm', in Moore and Schmitz, (eds) *Debating Development Discourse*, op. cit., pp. 30–51. The World Bank's annual *Development Reports*, in particular, came more and more to resemble flip charts for neo-liberal propaganda, a sad decline for what was once, for all its faults, a formidable intellectual machine.

'Development studies' as a substitute for 'development theory'?

In 1991 a group of left-inclined development theorists collaborated to produce an excellent volume, edited by Frans Schuurman, called *Beyond the Impasse: New Directions in Development Theory.*[60] The idea of an 'impasse' in critical development theory had been canvassed in various articles of the mid-1980s, including one by David Booth;[61] by 1991, however, Booth, in a leading contribution to the Schuurman volume, saw signs that the impasse was being overcome. In his view (strongly endorsed by Schuurman) empirical research had emancipated itself from the excessive generality, necessitarianism, teleology, class reductionism, dogmatism and other shortcomings of Marxist-influenced development theory, and in doing so had begun to show a potential for fresh theoretical initiatives. New theory would be sensitive to the great diversity of situations in the Third World, would refuse to reduce complex and locally specific gender and other relations to relations of class, and would allow for the possibility of 'room for manoeuvre' at the 'micro' and 'meso' levels of action, as well as the 'macro' level, which had been the focus of previous development theory — without, however, abandoning the inherited agenda of political economy. It might also, Booth hoped, succeed in combining the study of the cultural meanings subjectively attributed to things by people with the study of those same things from an external or objective standpoint, in a way not achieved before: and it would try to be relevant to the concerns of those engaged in practical development work.[62]

Booth's starting-point — the enormous expansion of field research under the aegis of 'development studies' — was, of course, valid; and if we add to this the no less impressive accumulation of social research not necessarily conceived of as 'development studies' — including social and economic history and gender studies in Third World countries — both the volume and the quality have clearly outstripped what went before, frequently revealing the somewhat shaky empirical foundations of previous 'grand theory' as well. And Booth's characterization of this work as mostly free from some of the vices of earlier development thinking — reductionism, excessive generality and the rest — and as being much more varied in its

[60] Frans J. Schuurman (ed.), *Beyond the Impasse: New Directions in Development Theory* (London: Zed Books, 1993).

[61] David Booth, 'Marxism and Development Sociology: Interpreting the Impasse', *World Development* 13/7, 1985, pp. 761–87; see also, among others, Peter Vandergeest and Frederick H. Buttell, 'Marx, Weber and Development Sociology: Beyond the Impasse', and Leslie Sklair, 'Transcending the Impasse: Metatheory, Theory, and Empirical Research in the Sociology of Underdevelopment', both in *World Development* 16/6, 1988, pp. 683–95 and 697–709.

[62] David Booth, 'Development Research: From Impasse to a New Agenda', in Schuurman (ed.), *Beyond the Impasse*, op. cit., pp. 49–76. This summary is necessarily crude, of course. Among various judicious reservations Booth points out the obvious problems of defining for whom and to what development theory should be relevant.

interests, was also accurate. Women, local-level activity, ethnicity, religion and culture, for example, which all tended to be secondary in the earlier literature, are often foregrounded in more recent work, which also tends to show more concern for detail, sets higher standards of proof and is in many ways intellectually refreshing.[63]

But Booth's idea that new development theory will emerge autogenetically from the accumulating volume and density of all this work, through some spontaneous fusion with the concerns of previous political economy, is a different matter. On the one hand, these 'mini-narratives' (if one may so call them, in contrast to the old 'big meta-narratives' which it is now fashionable to disclaim) have implicit higher-level theoretical presuppositions that need to be made explicit (microfoundations imply macrostructures, as much as the other way round), and it would be surprising if these were found to constitute, so to speak spontaneously, a new and better theory of development. And, even more crucially, the construction of a new theory of development is necessarily a political task, involving political choices about whom (what social forces) the theory is for, to accomplish what ends and in what contexts. Conflicting political commitments were, after all, what ultimately inspired the powerful theoretical debates within 'development theory' in the 1970s, and any worthwhile renewal of development theory now depends on a renewed clarification of political presuppositions and purposes as well.

To put it another way, what is striking about the way Booth and his colleagues conceive of the 'impasse' and its transcendence is that it is so idealist, i.e. the origins and the solution of the problem seem to lie in theory itself. With the exception of a page in Schuurman's 'Introduction', little reference is made in the book to the changes in the real world that have undercut the original development project.[64] There is unease about what 'development' now means, and the authors presumably agree with Schuurman's disclaimer that the goal cannot be 'one grand and glorious metatheory' (p. 32). But they do not confront the thought that, so long as collective socio-economic interests are supposed to be the products of the action of market forces rather than goals of strategic state action, the domain of 'development theory' is radically changed, if not abolished; that what is left is simply a world economy whose effects are overwhelmingly

[63] For an exemplary review of a great deal of this work see Frederick Cooper, Florencia E. Mallon, Steve J. Stern, Allen Isaacman and William Roseberry, *Confronting Historical Paradigms: Peasants, Labor, and the Capitalist World System in Africa and Latin America* (Madison: University of Wisconsin Press, 1993).

[64] The most explicit references are in Schuurman's 'Introduction', where he refers to 'the conviction that the world market is an overarching whole which cannot be approached using development policies oriented at the national level' and 'the growing recognition of differentiation within the Third World' (p. 10). There are also passing statements in the chapter by David Slater on 'The Political Meanings of Development' (e.g. 'Couched within a certain tradition one might even suggest that we are living the "end of development"' (p. 93)) but neither author pursues the apparent implications of these remarks.

determined by very powerful states and market actors, with at most minor modifications or delays brought about by the actions of lesser states, social movements, communities or whatever, which do not have significant military or market power. Certainly, this thought may be mistaken, or at least exaggerated; but in that case a new theory of development must at least begin by showing why.

To repeat what was said earlier, this is not to say that there is no further use for theory. But theory needs both a subject and an object, and the prerequisite of any new development theory that aims to be practical must surely be the analysis of the now deregulated global market and the social forces that dominate it, and then a definition of alternative social forces whose developmental needs cannot be met within this system, and which can be expected to struggle against it. Simply abjuring the alleged shortcomings of the theories that were constructed in the period of the collectively regulated world economy of legally sovereign states, and accumulating ever more detailed and subtle empirical analyses of local and particular experiences, will not of itself answer this need; for that world economy is, as Hegel put it, a form of life that has become old, and which theory cannot rejuvenate but only understand.

Eclecticism as usual in the development community?

The term 'development community' refers here to the network of people professionally concerned with development — the staff of 'donor' and recipient country development ministries, of multilateral 'aid' agencies, financial institutions and non-government organizations, and academic and non-academic consultants. It implies no disrespect to say that this community also constitutes an interest which has to adapt as best it can to constantly changing circumstances, rather like civil servants at the national level. There is a broad consensus about aims and possibilities, founded on development economics but honed by experience and the perspectives of other disciplines into a somewhat eclectic mixture capable of absorbing sometimes quite drastic changes in fashion or politics (witness, for example, the World Bank's successive accommodations, first to a mild touch of dependency, and then, within a decade, to neo-liberalism); and then, when the need passes, reverting back to a more centrist stance. How has the development community responded to the new situation?

As an example we may take John Toye's widely acclaimed study, *Dilemmas of Development*, which I have already cited. Toye's book is well known as a trenchant critique of the neo-liberal dogma which gained ascendancy in the World Bank and IMF in the 1980s, but it is also quite a revealing statement of his own position. For, besides criticizing the neo-liberals, Toye also criticizes the old political economy ('left wing', in his terminology) to which neo-liberalism was a reaction. Toye treats 'left-wing' political economists in rather general terms, and even lumps them together as the exponents of what he calls 'the standard left view' (for example, of the state (pp. 121–22)), even though it is sometimes hard to think of any individual theorist who has actually subscribed to the view he describes; but what is interesting is that Toye seems to be at least as hostile towards them as to the neo-liberals, and this draws attention to the fact that the ground Toye sees himself occupying is the sensible, reasonable, middle ground, in between these untenable extremes. What is the nature of this terrain?

The answer is not immediately obvious. Toye is a careful and penetrating critic of other theories, but the standpoint from which his criticism is made is not so clear. For instance, he explicitly subscribes to the following views, among others: 'global modernization' is 'inherently conflictual' because it is 'a human directed historical process' (p. 6); what is practicable and desirable is 'managed capitalism' (p. 10); in seeking to promote development we must avoid bringing preconceptions from outside, and see things through the eyes of poor people in the countries concerned (p. 40); and the recent economic retrogression in so much of the Third World 'appears to be a short-period interruption to a long period of buoyant growth' (p. 34).

Now, none of these ideas is self-evident. Why should human-directed historical processes be considered inherently conflictual, rather than inherently collaborative? What theory of history or human nature is involved here? On what grounds does it make sense to believe in 'managed capitalism' as an ideal, given capital's dramatically successful escape from management into the realm of the 'self-regulating' global market since the 1970s? On what grounds and in what circumstances are the perceptions of poor people to be respected, relative to other kinds of understanding? (What makes their thinking about economics important or valid, but not, for example, their ideas about the supernatural?) And from what theoretical standpoint does the retrogression of the last decade or more appear as a short-period interruption to growth, rather than as a long term, if not permanent, reversal?

Toye undoubtedly has answers to such questions. What is interesting is that he does not seem to feel the need to offer them; he does not defend his assumptions, eclectic and open to challenge as they are.[65] He attacks both the left and the right for being teleological, for theorizing development as 'the gradual movement to a single desired social state', calling instead for a 'modest', 'realistic' acceptance of diversity and of the ultimate unpredictability of events (pp. 14–16). This may sound sensible; it undoubtedly makes

[65] See David Moore: 'It might be...*apropos* to call the next phase [of development discourse]...one of pseudo-eclecticism. In it, theoretical and political jostling over such issues as sustainable development, state and/or market-led economic policies, and participation (or "democratic governance") will take up much of the public stage. The World Bank, the IMF and their lesser kin will appear to take pragmatic and consensual approaches to their hearts, much like in the 1970s, but will rest assured that their real neo-liberal agenda remains intact' ('Development Discourse as Hegemony', in Moore and Schmitz (eds), *Debating Development Discourse*, op. cit., p. 9).

it easier for the business of the development community to continue at a time when a more demanding canon of theoretical coherence would make it more difficult.

Or a more demanding conception of development as a goal. For it is also worth reflecting on what Toye puts in place of the 'single desired social state' (another straw man — who on the right or the left has really advocated this?) that he rejects as a 'teleological' approach to development: 'what most people would say mattered ultimately', he suggests, 'is the ending of large-scale poverty... sickness, ignorance and premature death, not to mention the violence, ugliness and despair of daily life' (p. 36). Apart from the fact that this seems no less 'teleological' than any other goal of development, where do these values come from? Who are these 'most people' whose authority is being appealed to here, and who no longer care about equality or democracy? Is this 'common sense', in whose name theory, left and right, is attacked, anything other than the tradition of Western charity?

Dependency theory in the 1990s

Dependency theory in its early sense of a general theory that sought to explain underdevelopment at the periphery as almost wholly the insuperably self-perpetuating effect of metropolitan capital probably has few remaining adherents. The NICs showed that structures of dependence might sometimes be overcome, while growing interdependence among even industrialized economies has made all dependence relative. However, dependency theory's focus on the many forms of acute dependence of small, open, ex-colonial economies on the powerful economic interests and states that dominate the financial and commodity markets in which they operate ('concrete situations of dependency') remains indisputably valid. In this primarily heuristic sense of the term, dependency theory has passed into the standard conceptual toolkit of most people who seriously study problems of development, and no longer needs or attracts much general exposition.[66] But it is interesting, all the same, to see how the changes that have

[66] An exemplary exposition of dependency thinking as a central dimension of the analysis of the impact of global market forces on the periphery is to be found in a long series of influential papers by Manfred Bienefeld, some of which have already been cited. Others include, notably, 'Dependency in the Eighties', *IDS Bulletin* 12/2, 1980, pp. 5–10; 'The International Context for National Development Strategies: Constraints and Opportunities in a Changing World', in Manfred Bienefeld and Godfrey Martin (eds), *The Struggle for Development: National Strategies in an International Context* (Chichester: Wiley, 1982); 'The Significance of the Newly Industrialising Countries for the Development Debate', *Studies in Political Economy* 25, Spring 1988, pp. 7–39; and 'Structural Adjustment: Debt Collection Device or Development Policy?', ADMT Series No. 5 (Institute of Comparative Culture, Sophia University, Japan, 1993). It is also worth noting that in its least threatening form (see note 24 above) dependency theory in Latin America played a valuable role in helping to legitimize somewhat more critical development research in the USA, as, for example, in the work of Peter Evans (*Dependent Development: The Alliance of Multinationals, the State and Local Capital in Brazil* [continued on page 32]

occurred in the real world have been seen by one of dependency theory's most famous Western exponents, Andre Gunder Frank, who has described the changes in his own thinking in an autobiographical essay published in 1991.[67]

Much of the charm of this memoir lies in its documentation of Frank's successive friendships and quarrels with other development theorists, and its idiosyncratic mix of (probably valid) claims to have been far in advance of general opinion on nearly every issue of development theory, and (no doubt equally valid) admissions to having been wrong (along with later general opinion, of course). What concerns us here, however, is what Frank finished up thinking. On dependency theory, his own summarized auto-critique reads as follows:

> 1) Real dependence exists, of course, and more than ever despite denials to the contrary. However, dependence 'theory' and policy never answered the question of how to eliminate real dependence and how to pursue the chimera of non- or in-dependent growth. 2) Dependence heterodoxy nonetheless maintained the orthodoxy that (under)development must refer to and be organised by and through (nation state) societies, countries or regions. However this orthodox tenet turns out to be wrong. 3) I turned orthodoxy on its head, but I maintained the essence of the thesis that economic-growth-through-capital-accumulation equals development... Therefore, I precluded any real alternative definitions, policy and praxis of 'development'. 4) In particular, this orthodoxy incoporated the patriarchal gender structure of society as a matter of course. However much I may personally have been against male chauvinism, I thereby prevented examination of this dimension of dependence. (p. 37)

Points (1) and (3) are the most telling criticisms made by sympathetic commentators on Frank's work at the time of its greatest currency.[68] Point (4) acknowledges a blind spot which Frank's partner and collaborator Marta Fuentes induced him to recognize. But it is point (2) that was the most significant for the subsequent course of Frank's thinking. What does it mean to say that development is not something for societies, countries or regions to organize?

As the 1970s progressed, Frank, exiled from Chile to England following the coup and the murder of his friend and early protector, President Allende, became convinced that not only was dependency theory devoid of any convincing alternative conception of development, but the whole idea of national development, which had been the *raison d'être* of development theory, was no longer tenable in the emerging conditions of a worldwide market. His historical work on the emergence of the capitalist world system led him

(Princeton: Princeton University Press, 1979)); or even David Becker and Richard Sklar's so-called 'post-imperialist' thesis, which asserts the validity of dependency theory's treatment of development as capitalist while appearing to be combating its allegedly erroneous ideas about imperialism (David G. Becker *et al.*, *Postimperialism: International Capitalism and Development in the Late Twentieth Century* (Boulder: Lynne Rienner, 1987).

[67] Andre Gunder Frank, 'The Underdevelopment of Development', op. cit., (note 21).

[68] See Chapter 2 below.

to see all dreams of alternative development paths pursued by particular countries or regions 'delinked' from that system as illusory (a conclusion fortified by the productive as well as moral bankruptcy of 'actually existing socialism'), while his work on the 'crisis' into which the world capitalist system had entered in the 1970s convinced him that for most of the Third World development within the system was also impossible. In particular, so long as Third World countries were paying on average about 6.5% of their gross national product (GNP) to service their debt, as he estimated was the case through much of the 1980s, there could only be 'development of underdevelopment', with '*dis*investment in productive infrastructure and human capiital and with the loss of competitiveness on the world market'.

The theoretical position to which this led Frank was that the only useful object of study is 'world development', which sets the limits to whatever normative goals it makes sense to try to pursue, and that the only useful agents capable of pursuing such goals are 'particular groups or classes' (p. 54). Such goals, even if they can be achieved by such groups or classes, will be relative to the way world development currently affects the part of the globe they live in, a development that has been going on throughout recorded history:

> I now find the *same continuing world system*, including its center-periphery structure, hegemony-rivalry competition, and cyclical ups and downs has been evolving (developing?) for five thousand years at least... In this world system, sectors, regions and peoples *temporarily and cyclically* assume leading and hegemonic central (core) positions of social and technological 'development'. They then have to cede their pride of place to new ones who replace them. Usually this happens after a long interregnum of crisis in the system. During this time of crisis, there is intense competition for leadership and hegemony. The central core has moved around the globe in a predominantly westerly direction. With some zig-zags, the central core has passed through Asia, East (China), Central (Mongolia), South (India) and West (Iran, Mesopotamia, Egypt, Turkey)... Then the core passed on to Southern and Western Europe and Britain, via the Atlantic to North America, and now across it and the Pacific towards Japan. Who knows, perhaps one day it will pass all the way around the world to China. (pp. 56–57)

Development theory based on any idea of 'autonomous' national development, on any conception of 'de-linking', is, therefore, an illusion (p. 58). What is needed is 'a more rounded, dynamic and all-encompassing supply and demand side economics to analyse, if not to guide, *world* economic and technological development'. The significance of the words 'if not to guide' is not made clear: perhaps Frank is still faintly agnostic on the possibility of some form of world government emerging? But, for any particular country, region, sector, group or individual, he is very specific: having a development 'policy' simply means finding 'one or more niches in which to carve out a temporary position of "comparative" monopoly advantage in the international division of labour' (p. 57). 'But for much of Africa, and probably also Bangladesh, the Brazilian north-east, Central America, the North

American rustbelt and much of the former socialist world, the prospect is only of marginalization and decline. Most development for one group... comes at the expense of anti-development for others. They are condemned to dualistic marginalisation and/or to underdevelopment of development. That is what real world development really means'(pp. 58–60). And, since all existing models of development are inadequate, Frank pins his hopes only on radical democratization, based on the emerging strength of the hitherto neglected social groups, and especially women.

The pessimism of Frank's conclusion does not ultimately rest on theoretical grounds, however (any more than does the optimism of the *Washington Post* cited earlier). For example, at one time e was an exponent of regional trading blocs committed to the defence of shared political and social values — the sort of thing of which the Social Chapter of the Maastricht Treaty is a pale reflection — but he seems to have abandoned this idea without giving any reason for no longer believing in its potential.[69] His loss of faith in the power of states to achieve anything in present conditions is not argued for either, even if it is fairly evident that the states of the countries that most need effective development policies are those least likely to be able to produce or implement one.

It is to this point, of course, that interpreting the success of the NICs is so crucially relevant. At one time their success was attributed by neo-liberals to the virtues of *laissez faire*, until the work of Hamilton, White, Amsden, Wade and others showed incontrovertibly that, if anything, the NICs' experience demonstrated the precise opposite, i.e. the necessity for forceful, systematic and sustained economic intervention by a strong, centralized state pursuing a coherent long-term development strategy.[70] This conclusion has now been swallowed, albeit with some difficulty, by the World Bank;[71] but its implications for 'development theory' have still to be fully digested in 'mainstream' circles.

To see this, it is worth looking briefly at the work of the American political scientist, Stephan Haggard. Haggard does not see himself as a dependency theorist, but he did set out to see what conclusions should be drawn from the experience of the NICs, using the methods of US political

[69] Even if the European Union's (EU's) record so far is not inspiring, in terms of its constitution, and especially relative to the North American Free Trade Agreement, it is a significant advance in supranational government with, in principle, the potential not merely to protect people from the negative impact of the market on their region of the world, but also — through the weight of the EU in the world economy — to affect the course of world economic development in the sort of way Frank now seems to think impossible.

[70] See Clive Hamilton, *Capitalist Industrialisation in Korea* (Boulder: Westview, 1986); Gordon White, *Developmental States in East Asia* (New York: St Martin's Press, 1987); Alice Amsden, *Asia's Next Giant: South Korea and Late Industrialization* (New York: Oxford University Press, 1989); and Robert Wade, *Governing the Market: Economic Theory and the Role of Government in East Asian Industrialization* (Princeton: Princeton University Press, 1990).

[71] See *The East Asian Miracle: Economic Growth and Public Policy* (New York: Oxford University Press for the World Bank, 1993).

science.[72] Through a comparative study of the recent history of South Korea, Taiwan, Singapore, Hong Kong, Mexico and Brazil, he identified four main causes for their varying degrees of success: first, external pressure for change, in the form of price shocks, reinforced in several cases by great-power influence and leverage; second, favourable constellations of political interests, especially a weakened agricultural élite; third, a state with the capacity to formulate coherent and centrally implemented policies, with suitable policy instruments (financial institutions, fiscal and regulatory regimes, etc.), and a strong degree of political insulation from even economically powerful interests; and fourth, technical competence, 'policy-relevant knowledge'. Haggard analysed how each of these conditions, which were present to different extents and in very different forms in the six cases studied, interacted with their varying geographic and resource endowments to make possible the dramatic adjustments involved in the rapid transition to international manufacturing competitiveness.

His conclusions were cautious, but in general very clear: all of these conditions seemed to have been important, if not indispensable, and perhaps above all the ombination of consistent policy (even in the case of Hong Kong, which pursued a policy of almost complete *laissez faire*) with strong insulation from political pressure from affected sections of society, including powerful agricultural and business interests. While authoritarian rule was clearly not a sufficient condition for competitiveness, Haggard concluded, it may often be a necessary one.

The substantive implication of Haggard's findings seems to be that not many underdeveloped countries are likely to meet the necessary conditions in the foreseeable future. Furthermore, Haggard did not analyse the distinctiveness of the international context at the time of these countries' decisive advances, i.e. how far their development successes depended on such factors as buoyant conditions of world demand, or the exceptionally large capital flows and access to the US market made available by the USA to Korea and Taiwan — factors very unlikely to replicated elsewhere.

There is, finally, a further theoretical point that needs to be made. Haggard's study offered no grounds for trying to answer the question of how any other country might ever put itself in the position to replicate the NICs' conditions for successful industrialization. His comparative analysis did not extend to the specific histories that produced the relative sophistication of these countries' pre-existing markets, technology and capital stock, the competence of their entrepreneurial classes and state technocracies, the relative autonomy of their state apparatuses and the relative political

[72] Besides his useful extension of the analysis to include two Latin American 'NICs', Haggard has had a greater impact on US political scientists than Amsden or the writers of the 'Sussex school', which is a further reason for focusing particularly on his work here. Haggard also dissociates himself from the ritual attacks on dependency theory that seem to have become almost a test of soundness for many US political scientists: see Stephan Haggard, *Pathways From the Periphery: The Politics of Growth in the Newly Industrializing Countries* (Ithaca: Cornell University Press, 1990), p. 19.

weakness of their landed and working classes. But for the analysis of the NICs to contribute to a new theory of development these histories are evidently crucial: can we say anything at all general about the processes that each of these countries had to pass through to produce the specific mix of conditions that made its breakthrough possible? What cultural characteristics permitted the formation of the necessary consensus within its state apparatus about the development strategies it adopted, and the necessary technical understanding to implement them consistently and effectively? To answer this kind of question, however, we are forced out of what Bayart calls 'weak ahistorical comparativism' into the study of comparative historical dynamics,[73] i.e. into the kind of enquiry for which there is still no serious alternative to some of the central theoretical ideas of Marx.

Rational choice and development

One of the most influential reactions — at least in the USA — to the end of 'development theory's' *raison d'être* has been to try to build a new political economy of development in which the key to economic performance is seen as institutions that can be analysed in terms of rational choice theory.

This response makes sense in several ways. First, it involves going back to the problem raised by Myrdal and his colleagues in the 1960s — how to incorporate the obstacles posed by political and social phenomena into the analysis of economic development — but this time in a way that explicitly tries to stay within the assumptions of neo-classical or marginalist economics, that is unlike the neo-Weberians of the modernization school (not to mention Marxists or neo-Marxists), the 'new institutionalism' is supposed to rest on the assumption of rational individuals maximizing their utilities and nothing more, and should (in its most optimistic version, at least) be capable of being integrated with economics and modelled mathematically. Given the recent ascendancy of neo-classical economics in the 'development community', this has an air of political realism about it (and maybe — some people evidently hope — it can also endow political science with some of the economists' famous rigour).

Second, it holds out the prospect of dispensing with the Marxist phenomenology of classes and relations of production and other unclean entities, whose relevance to the problems of development cannot always be denied; in the choice-theoretic discourse all of these are reducible to special cases of a very small stock of extremely general concepts, such as institutions, organizations and their principals and agents.

Only some comments on the foundations of this school of thought will be made here; a somewhat fuller discussion is offered in Chapter 4 below.

[73] Jean-François Bayart, 'Finishing with the Idea of the Third World: The Concept of the Political Trajectory', in James Manor (ed.) *Rethinking Third World Politics* (London: Longman, 1991), pp. 53–54.

The central idea of the 'new institutionalism' or 'new political economy' is that what makes for an efficient economy is a set of institutions that permit individuals to benefit personally from doing what will also serve the (material) interests of society as a whole. Thus, for instance, a system of land tenure that allows tenants to keep for themselves a significant part of any expanded output they produce through allocating extra resources of capital or effort to its production is more economically efficient than one which does not. This reasoning can be applied to taxation, the organization of central or local government, education, banking, marketing — in effect, to any social arrangements (even marriage law and custom). Conversely it is often possible to see, retrospectively, that the institutional structure has provided incentives for individuals to do things which were inimical to development; while prospectively it is often possible to imagine or even design institutional arrangements that will improve the social returns to the economic activity of individuals (which is, roughly, what management consultants are supposed to be concerned with when they are hired by the state).

But in the absence of further arguments this does not amount to a theory of development, or even a significant addition to existing theories, for a very simple but fundamental reason: we cannot explain in terms of the 'paradigm' how any particular set of institutions that existed in the past or exist today in a given country came into existence. For that we have to resort to a much wider, looser theory of social change of precisely the kind that most exponents of public choice theory are trying to dispense with. Douglass North, who can fairly claim to be the chief inspirer of the idea that choice-theoretic thinking can yield a new theory of development, is strikingly ambivalent on this point. On the one hand, he is clear that

> Neoclassical economics...may account very well for the performance of an economy at a moment of time or, with comparative statistics, contrasts in the performance of an economy over time; but it does not and cannot explain the dynamics of change. The major source of changes in an economy over time is structural change in the parameters held constant by the economist — technology, population, property rights, and government control over resources. Changes in political-economic organisation and its consequent effects are basic to theorising about all these sources of structural change.[74]

And he even adds that

> The Marxian theory is the most powerful of the existing statements of secular change precisely because it includes all of the elements left out of the neoclassical framework: institutions, property rights, the state, and ideology[75]

— even though he thinks the Marxian theory has various faults and is, at best, a special case of change and unacceptable as a general theory.

[74] Douglass C. North, *Structure and Change in Economic History* (New York: W.W. Norton, 1981), p. 57.
[75] Ibid., p. 61.

On the other hand, he thinks that changes in institutions (which on his definition include property rights and the state) can be incorporated into neo-classical analysis through the concept of a 'path' of change that flows from the interaction between the existing institutions and the organizations that individuals form to maximize their utilities within the framework of incentives offered by the institutions. This interaction leads to modifications of the institutions over time, typically in an incremental way that tends to reinforce whatever tendency the original institutions had to be socially efficient or inefficient; this is roughly what is meant by the 'path dependency' of a social system in this parlance. Once an adequate inventory of a country's existing institutions and organizations has been made, then, it should be possible to infer the development path that it will follow as a result of their interaction; even the modifications that are likely to be made to the institutions, and hence the subsequent changes in organizational behaviour, should be deducible. In other words, a general development theory of the kind Marx proposed, but capable of being integrated with neo-classical economics, should be within our grasp.

There are, however, three basic difficulties with this idea. First, the concept of 'ideology' has to be expanded to deal with everything that cannot be included within the assumptions of choice theory. North is quite aware of the fact that he must rely on 'ideology' to deal with such problems as historically decisive instances of collective behaviour which on strict individual utility-maximizing assumptions should not have occurred because of the 'free-rider problem' (i.e. they should not have been possible because it would have been in everyone's best individual interest to leave it to others to run the risks and pay the price involved); and so he calls for a theory of ideology to complement the theory of institutions.[76]

But it is not clear that he recognizes how fatal this is to his project. So much of the problem of understanding social change is understanding what motivates collective action, and the results of centuries of study and reflection suggest very strongly that there are not going to be any general, or any simple, answers. Any interesting answers have always been specific to historically well-studied places and times (the French Revolution, the nineteenth century labour movement in Europe), and have involved complex long-term and short-term interactions between individuals, groups, cultural practices and institutions of specific kinds (churches, constitutions, professions, communities, armies) — in short, they almost always involve the social whole, including many of the very things ('institutions') which for North's project must be kept analytically separate from 'ideology'.

A second difficulty is the project's reductionism. For instance, the idea that developmentally significant change may be understood as being the result of the interaction between existing institutions and the organizations formed to achieve whatever goals the institutional structure makes possible and attractive is, obviously, a very general statement about the sort of

[76] Ibid., Chapter 5.

relationship that Marx postulated between classes and property rights. Presumably the advantage of reformulating it in these terms is that it brings out the general characteristics, which any such hypothesis needs to have, in a way that does not prejudge what it will actually state. The difficulty is, however, that whatever plausibility the general statement has comes from the particular case, not from the abstract one, which looks like a tautology. In effect, North is stating that any theory that is to replace Marx's needs to be of this general character. But, in so far as this is true, it is not new, and, in so far as it is new, it is not obviously true. In so far as it is true, few Marxists today would disagree that we need a theory of Marx's kind, but better; the problem is only to have the learning and analytic power to come up with one. But, if the suggestion is that we need a theory that can be formulated entirely within the assumptions of rational choice theory, its truth is by no means obvious. That idea comes from a preference for rational choice theory based on metatheoretical grounds, not because it has been shown to be suitable for handling large-scale and long-term social change, since (as North himself notes) this is not the case.

Third, there is a closely related tendency, even in North's thinking (which is in general sensitive to such problems), to argue that, because some aspect of observed reality can be modelled, that aspect is the determinative or key one. For instance, institutions are very broadly defined, in this literature, as systems of rules or norms constraining behaviour, which means that virtually all persisting social relations can be represented as institutions.[77] But then the claim that institutions 'are the underlying determinant of the long-run performance of economies' becomes an unhelpful truism — i.e. the pattern of social relations determines economic performance.[78] Yet when North makes this claim it does not feel like a truism. This is because by this time he has built a rational choice model of certain aspects of institutions having to do with property rights and their related transaction costs, incentive structures and so on, which have obvious economic implications.

But what about the determinative effects of all the other aspects of all the other kinds of institutions not susceptible of being modelled in this way? What about the effects of the passion aroused in religious movements, or the conservatism, loyalty, discipline, etc., embodied in cultural norms, or the reforming or revolutionary zeal generated by class or national feeling, all of which seem to have played no less crucial parts in determining economic performance at one time or another in history? They can be brought back in only by accepting that the claim that institutions are the 'underlying' determinant is true by definition. It is plausible that, other things being held constant, property laws will have important effects on

[77] 'Institutions are the rules of the game in a society, or, more formally, are the humanly devised constraints that shape human interaction' (North, *Institutional Change and Economic Performance* (Cambridge: Cambridge University Press, 1990), p. 3).
[78] Ibid., p. 107.

economic performance. But the whole difficulty of understanding development, as North frequently acknowledges, is that other things do not stay constant but continually interact with property rights and all other kinds of social relations in ways that cannot be comprised within any model as simple and one-sided as those of rational choice theorizing.

Thus, while rational choice undoubtedly has valuable contributions to make to specific issues in development — the work of Samuel Popkin on peasant farmers' behaviour is an excellent example — it does not point the way towards a new development theory for our times.

'Rethinking Third World politics'

Assuming, that is, that we are still concerned with development. The response of some academics has been, on the contrary, to frankly abandon it. Thus James Manor writes in the introduction to *Rethinking Third World Politics* that he and his fellow contributors

> are not interested...in influencing the countries we are studying. We do not even intend these studies to advance the cause of democracy in the Third World, even though we wish it well. We believe that that cause is best served by studies which provide the most sophisticated possible understanding of how things actually work in Third World political systems.[79]

Leaving aside the question of how research that appears to deny its own value commitments could be sophisticated, it is interesting to see what is involved in abandoning any policy concerns, including any commitment to 'development', by examining the work of the most fertile and wide-ranging of Manor's contributors, Jean-François Bayart.[80]

In his chapter entitled 'Finishing with the Idea of the Third World: The Concept of the Political Trajectory', Bayart outlines his concept of 'historicity', the idea that politics must always be understood as a moment in a complex and very long-term story. This story can be understood, Bayart suggests, in three possible ways: as the story of a 'civilization' (in Braudel's sense), as the story of a system of inequalities (caste, class, age, etc.) or as the story of a culture — or as a combination of these. Out of people's experience of this past, or these pasts — pasts which, Bayart stresses,

[79] James Manor (ed.), *Rethinking Third World Politics* (op. cit.), p. 2.

[80] Manor's statement was doubly odd given that only five of the twelve contributions to the book actually conformed to it. One contributor, Garreton, was explicitly concerned with the conditions for democracy in Latin America, and four (Medard, Crook, Clapham and Sandschneider) were concerned with familiar development themes. Two more (Samudavanija and Hawthorn) argued, in effect, that Third World governments have generally been less interested in economic development and more interested in securing authority, defining national identity and so on than Western students of development would like: which is true, although not something that has entirely escaped the attention of students of development hitherto.

comprise external influences as well as forces endogenous to the country or region under study — they have constructed various 'discursive genres', in terms of which politics are understood: examples he gives include such widely differing genres of discourse as the British system of government (a discourse about representation, civil liberties, etc.), Islamic thought, and the 'world of the invisible' (the occult, witchcraft, etc.). People think in terms of these discourses, and politics are constrained by them while at the same time involving contestations between them. The analysis of politics must therefore, according to Bayart, try to link 'the collective work of the production of the state to the subjective interiority of its actors' (p. 68) by studying both the long-term historicity of a people, through which their political institutions have evolved, and the discourses through which these institutions are participated in and understood today.

These statements are programmatic. To assess their practical significance we can take Bayart's major study, *The State in Africa*, as a case in point.[81] In this book his stated purpose is to understand the 'historicity' of the African post-colonial state. He wants to transcend the shorter-term theories of the 'developmentalists', the 'modernization' school, the dependency school, 'the Marxists' and various others — all of whom share, in his view, more or less the same basic error, i.e. they ignore, or radically misunderstand, Africa's 'true historicity' (p. 5).

In contrast, Bayart offers a '*longue durée*' view of Africans as having over the centuries always been subordinate players in relation to the outside world, but players none the less, always engaged in a process of 'extraversion', in which they have sought to draw on resources or alliances available in the external environment in furtherance of their continuing internal competitions and conflicts. What we have now is only the latest version of this role, although now enacted on wider stages than in earlier times — in the ex-colonial territories, and in a number of regional 'power-spaces' (West, East and Southern Africa, polarized around Nigeria–Zaïre, Kenya and South Africa respectively). What is now going on, Bayart argues, is the construction of new historic blocs, 'rhizomatically' linked to the underlying societies (i.e. like shoots from a tuber) and clustering around the state, and actually combining elements that earlier theorists have tended to see as mutually exclusive and opposed to each other: traditional and modern élites, local and central élites, chiefs and civil servants, state and private-sector élites, etc. In Bayart's view, ethnicity, class and the rest are all interlinked in a 'reciprocal assimilation of élites', as the members of these élites collaborate with each other to profit as best they can from their dealings with the world outside.

And so what earlier theorists saw as deformations or aberrations appear in Bayart's optic as more or less normal, and in truth functional. Even a deeply corrupt state can be seen as an integrative force; even military coups can be understood as modes of intervention to cool out élite competition

[81] Jean-François Bayart, *The State in Africa: The Politics of the Belly* (London: Longman, 1993).

which has become out of control and destabilizing (p. 154); even structural adjustment programmes may be seen as removing spoils from the control of parts of the historic bloc that the president might otherwise not be able to dominate adequately (pp. 225-26). Bayart does not paint the post-colonial African state in rosy colours: the sick brutality of a Bokassa in the Central African Republic, a Touré in Guinea or an Eyedama in Togo is spelled out. But this kind of sadistic violence is seen as exceptional, while the colossal predations of less vicious rulers appear as essentially inevitable, inscribed in the continent's 'historicity'.

As for the African masses, Bayart frequently asserts that they are not passive victims of external forces, that they make their own history; but the actual role he shows them playing is circumscribed so closely by their lack of capacity to act for themselves and by their desperate struggle for survival that they much more often seem complicit in the trends he describes. At one point (p. 259) he even presents refugees as a kind of heroes ('escapees'), demonstrating the limitations of the power of state power-holders by exercising the 'exit option'; and yet in their very act of leaving Bayart also seems to make them responsible for the shortcomings of the situation they leave behind — their absence contributes to the weakness of civil society, the lack of capital accumulation, and hence to the existence of only a thieving, rather than a productive, ruling élite!

In the same passage, it is true, Bayart also characterizes refugees as 'unfortunate crowds hurrying towards frontiers and vegetating in reception camps'. But what, from the standpoint of the *longue durée*, is unfortunate about these crowds? For there is notable inconsistency in Bayart's account, in its oscillation between a sort of gruff realism about the post-colonial state and moral discomfort. For, where the modernization school expected the African élites to be modernizing and good, Bayart expects them to be what they are, interested in power, wealth and status at more or less any cost. His standpoint might seem Hegelian: history unfolds according to the cunning of reason, so that it makes no sense to shed tears for history's victims. To do so is inconsistent, and furthermore empty, since there is no way to intervene. But, unlike Hegel, Bayart does not subscribe to an 'objective idealism'. In his concept of history there is no higher purpose which people's suffering serves.

And this is what it means to try to study the Third World without any commitment. The work of those committed to 'development' had faults, but thanks to this commitment they all had some idea — however imperfect — of who they were writing for, and who might act in the light of what they wrote. Bayart's intended readers, on the other hand, seem to be ultimately just 'Africanists', capable of getting their minds round Africa's 'historicity', but with neither the power nor the wish to act historically. As with Toye's stance, this may have an air of being more 'realistic' than the stance of Marxists, dependency theorists or modernizers, but what does this amount to?

Bayart has evident affinities with post-structuralist discourse, according to which we can never know reality but can only make a variety of statements about it with varying degrees and kinds of usefulness. Among social scientists a frequent symptom of this idea is to lay stress on the complexity of everything and the way no one formulation ever fully captures it, a 'distancing effect' that certainly seems to play a part in Bayart's work. But, even within that discourse, something eventually is said, a choice of statements is made, a general account emerges. And then, it is fair to ask, from what standpoint is Bayart's ultimately quietist picture drawn? And for whom is it painted, if not for the aforementioned kleptocrats, whom it does not exactly celebrate, but does not condemn either?

Consistently, for someone uncommitted to any concept of development, Bayart makes a resolute separation of politics from economics and says virtually nothing about the relation between them. In his account of Africa, what matters is only how economic resources are appropriated to service the endless cycle of the reciprocal assimilation of élites. If at the end of the twentieth century many African countries are destined to suffer desertification, famine, crime and warlordism, or to undergo recolonization as vast refugee camps, is it of great importance in the *longue durée*? In practice, Bayart has been an active spokesman for African interests in French public debate. The stance outside or above the fray that he explicitly adopts in his thesis on historicity contradicts this, and it is this contradiction that repeatedly surfaces in these texts.

In conclusion: development, or the fate of the ancients?

These sketches of a few selected currents in contemporary writing about development (or, in the case of Manor and Bayart, in reaction against development) are, of course, subjective and partial. Their point is to raise the question of what 'development theory' was and has become, and above all to try to clarify what seems to me to be at stake: namely, the urgent need to revive development theory, not as a branch of policy-orientated social science within the parameters of an unquestioned capitalist world order, but as a field of critical enquiry about the contemporary dynamics of that order itself, with imperative policy implications for the survival of civilized and decent life, and not just in the ex-colonial countries.

Since the late 1960s, the debate about 'development theory' has in fact been more and more clearly about the theory of global development that each one presupposed, although the participants have all too often not recognized, or not acknowledged, that this was the issue at stake. Today it has to be frankly confronted: what do 'the universal development of the productive forces' and a truly global relation of supply and demand, which the OECD governments and the international financial institutions have been labouring for almost two decades to realize, now imply for any individual project of 'development'? For whom, contemplating what goals and by what means, can a useful 'development theory' be constructed?

The scale on which these questions seem to oblige us to think is painfully vast, and may seem almost to threaten incoherence; but, if it was not impossible to have a theory of capitalism on a national scale, why should it be impossible to have one of capitalism on a global scale? The theories of Hegel or Marx (or Weber or Fukuyama, for that matter) are not incoherent, but just very large-scale and necessarily full of selective simpifications, speculative elements, debatable assumptions and 'middle-level' problems of all kinds. What is really incoherent is a 'devlopment theory' that does not rest explicitly on as clear a general theory of world history, and of world capitalism in particular, as it is possible to have.

Such a theory must, evidently, indicate what is and is not possible for various potential actors, just as the Keynesian theory of global capitalism did at the birth of 'development theory'. On the basis of such a general theory, new development theories at a lower level of abstraction can then be formulated. These may be for states, for groups of states organized in regional or other organizations or for non-state agents of various kinds. The goals of development envisaged by these theories will depend on the actors for whom they are formulated and the scope for change that the theorist's preferred theory of world capitalism suggests exists for them. If, as I fear, it seems that not much scope for change exists — especially for small, severely underdeveloped countries — without a radical resubordination of capital to democratic control, development theory will have also to be about this, and agents capable of undertaking it.

This abstract conclusion seems to me preferable, in spite of its abstraction, to trying to breathe life back into any kind of 'development theory' whose illusory appearance of concreteness and 'practicality' depends on averting one's gaze from its lack of adequate foundations.

2

Underdevelopment & Dependency: Critical Notes

> This concept [dependence] fails to grasp the real nature of the process of underdevelopment. The immediate explanation for this is the theoretical framework employed...eclectic combination of orthodox economic theory and revolutionary phraseology.[1]

It is becoming clear that 'underdevelopment' and 'dependency' theory (UDT) is no longer serviceable and must now be transcended. The evidence for this is: (i) theoretical repetition and stagnation in the literature on UDT; (ii) the existence of fundamental problems of analysis which UDT cannot solve, or even formulate, and central problems of development strategy which are linked with these, and about which UDT is either silent or ambiguous; and (iii) an evident lack of practical impact in favour of the popular forces in the struggles in Third World countries, but on the contrary, a marked tendency for the underdevelopment/dependency (UD) 'perspective' to be co-opted by developmentalists allied to international capital. This chapter is an attempt to clarify all this, with the help of several recent sympathetic studies of the genesis of UDT and its variants, and to suggest what needs to be done about it.

UDT and its problems

By UDT I mean the following general theses:

1. The social, economic and political conditions prevailing throughout today's 'Third World' ('less developed countries' in United Nations (UN) usage) are not due to the persistence of an 'original' (undeveloped or 'untouched') state of affairs, but are the results of the same world-historical process in which the 'First World' ('developed market economies') became 'developed'; the development of the latter involved a closely associated

* First published in *Journal of Contemporary Asia* 7/1, 1977, pp. 92–107.

[1] Geoffrey Kay, *Development and Underdevelopment: A Marxist Analysis* (London: Macmillan, 1975), p. 103.

course of development for the former, a process of subordinate development or underdevelopment.

2. The prime mover in this combined process was capital seeking profits, i.e. seeking opportunities to accumulate capital — specifically, capitalist merchants, capitalist bankers, capitalist insurers, etc., and finally capitalist manufacturers.

3. Their activities involved accumulating capital where this could be done cheaply, and investing it where the return to investment was highest, and this gave rise to a process of surplus removal from some parts of the world to others, perpetuating and rigidifying in new ways the low levels of productivity in the areas from which the surplus was taken, and also a structuring of these economies so as to subordinate them systematically to the structures of the economies where capital was being accumulated. This expressed itself in the 'external orientation' of the subordinate economies (export of primary commodities, import of all manufactures); in monoculture; in dependent industrialization (dependent on external forces — demand, strategic division of labour, technology, etc. — for its dynamics and structuration).

4. Secondary structural consequences of this served to reproduce the process and constantly block local initiatives to pursue an autonomous development path; e.g. the low incomes of the majority due to the creation of surplus labour and marginalization imply a generally small domestic market; highly unequal income distribution implies a narrow import-orientated consumer demand; etc.

5. The corresponding emergence and formation of social classes at the capitalist periphery with interests in common with the bourgeoisie of the metropoles made possible the development of colonial, neo-colonial and semi-colonial states representing successive types of such alliances.

6. The term 'underdevelopment' refers to these self-perpetuating processes, to these self-reproducing structures and to their results. The term 'dependency' is sometimes used to refer to exactly the same things, and sometimes more specifically to refer to the non-autonomous nature of the laws or tendencies governing change in the social formations of the periphery. In spite of disagreements between the users of the two terms, their differences seem less important than their extensive points of general agreement.[2]

Of course this sort of outline misses part of the strength of UDT, which is to have stimulated the empirical study of institutional and structural mechanisms of underdevelopment such as multinational corporations (MNCs), income distribution, fiscal and taxation policies, capital expenditure and aid programmes, etc. But I think its main thrust is roughly as sketched above. And in the context of the early 1960s, when UDT emerged

[2] See F.H. Cardoso, 'Imperialism and Dependency in Latin America' in F. Bonilla and R. Girling (eds), *Structures of Dependency* (Nairobi/Palo Alto: Nairobi Bookstore, 1973), pp. 15–16.

as a militant critique of the ruling ideas of developmentalism, its thrust was unquestionably a progressive one. O'Brien's comment that 'the international economy is likely to prove a more seminal starting point for understanding development and underdevelopment than, e.g., traditional and modern society, stages of growth, achievement motivation, or even Myrdal's causally inter-related conditions of a social system' is an understatement which reminds us of some of the intellectual deserts from which UDT rescued us, along with 'integration', 'nation-building', 'political development', 'modernization' and much else. These were arid formulations because they started out from the premiss (more or less inarticulate) that underdevelopment was a 'primal' or 'original' condition and proceeded to ignore the historical and 'transitive' relationships between development and underdevelopment, and hence the whole phenomenon of imperialism. They nearly all ignored the existence of social classes, and treated the state as an instrument of popular will, or at least of the public interest as interpreted by a minority on behalf of the public. They had, in fact, a highly ideological character, as we can now easily see in the light cast by UDT. And some of them embodied this in the particularly stultifying forms of abstracted empiricism and 'grand theory'. In fact, all these features of bourgeois development theory have been made apparent by UDT. The question is, however, whether UDT is free from all of them itself.

As for UDT, then, it is true that the charge that it has tended towards theoretical repetition and stagnation is necessarily subjective; and it is probably more accurate to say that UDT was characterized from the first by a proliferation of attempts to formulate characterizations, models and analyses of underdevelopment or dependency in general.[3] It is significant that the few attempts at systematic theorization — for instance, by Amin or Szentes — have been by economists and have not really systematized the kind of total perspective — social, political, ideological and economic — implied in the more general formulations through which UDT is chiefly known. As I will indicate below, this reflects the genesis of UDT as a criticism of bourgeois development theory, which, while radical in intention, really remains within its problematic, so that the systematization of UDT could only produce a revision of bourgeois development theory, and an ultimately non-radical one at that. The sense of theoretical repetition and stagnation which the UDT literature eventually conveys (as O'Brien says, 'was it really necessary to write so many millions of words to establish just this perspective?') seems to me, then, to be due to the necessity of staying at the level of a general 'perspective' if the fundamental difficulties inherent in that perspective are not to be immediately apparent.[4]

[3] See, for instance, William C. Smith Jr, 'A Dependency Bibliography' in Bonilla and Girling, ibid.

[4] Philip J. O'Brien, 'A Critique of Latin American Theories of Dependency', in I. Oxaal, T. Barnett and D. Booth (eds), *Beyond the Sociology of Development* (London: Routledge and Kegan Paul, 1975), p. 25.

The sort of difficulties involved include the following (the list is not very ordered, let alone comprehensive):

The meaning of 'development'

This is obscure in UDT. In so far as it implies a concept of development it is evidently that of the capitalist development experienced by the capitalist 'metropoles'. But if the study of underdevelopment suggests that the underdeveloped countries in 1977 cannot follow a development path at all similar to that of the 'metropoles', it is a matter of great consequence if the analysis of how they became and are kept underdeveloped does not itself disclose what 'development' can or should mean for them, or how it is to be accomplished. The fact is that this does not seem to be so, and this is shown by the great variety of practical-political positions arrived at (often somewhat independently of their theoretical formulations) by different UD theorists; though UD theorists actually often adopt very *vague* practical-political positions, if any, and largely confine themselves to interpreting and explaining what has happened in the past.

Exploitation

Much UDT is unclear whether the masses in the underdeveloped countries suffer from exploitation or not, and if so, to what extent. Most UD theorists imply, at least, that the masses of the Third World are exploited as well as oppressed, but do not make it clear how. The issue was posed in the debate between Charles Bettelheim and Arghiri Emmanuel over Emmanuel's *Unequal Exchange* and is raised again in a sharp way by Kay, even though his formulation is not perhaps as clear as one would like ('capitalism has created underdevelopment not simply because it has exploited the underdeveloped countries but because it has not exploited them enough').[5]

Primitive concepts

It is hard to deny that UDT is, to say the least, very broad-gauged. The centre–periphery metaphor is seldom, if ever, replaced by a concrete typology of centres and peripheries. The same is true of 'dependence'. It is sometimes said (e.g. by Cardoso) that there are 'different forms' of dependence but, unless these are identified and the differences theoretically explained, the question arises whether this perspective is not a fairly arbitrary way of sensitizing us to one set of relationships at the cost of anaesthetizing us to others — i.e. sensitizing us to only one dimension of a more complex interdependency.[6] And it is not really an accident that these simplistic pairings, developed/underdeveloped, centre/periphery, dominant/dependent

[5] See C. Bettelheim in Appendices I and III to A. Emmanuel, *Unequal Exchange* (New York: Monthly Review Press, 1972), pp. 300–07 and 352–55, and Emmanuel, ibid., Appendix IV, pp. 380–83; and Kay, *Development and Underdevelopment*, op. cit., p. 55.

[6] D. Booth, 'Andre Gunder Frank: An Introduction and an Appreciation', in Oxaal *et al.* (eds), *Beyond the Sociology of Development*, op. cit., pp. 75–76.

resemble those of bourgeois development theory (traditional/modern, rich/poor, advanced/backward, etc.); they are basically polemical inversions of them. But the result is a very serious weakness: the empirical meaning of 'underdeveloped' has to be pretty slight if it must embrace India and Brazil as well as Haiti and Tanzania. It is also clear that Portugal, Britain, the USA and Japan (not to mention the USSR) are 'developed' in different ways that are seldom if ever clarified in UDT, important though this is. (Before 1974, for example, it was conventional wisdom for some UD theorists that a neo-colonial solution in Africa was 'impossible' for Portugal because of her own underdevelopment; it is clear that the concepts these words denote are far too weak for making the theoretical distinctions needed to analyse the important political questions involved here).

'Economism' and 'mechanicism'

UDT tends to be economistic in the sense that social classes, the state, politics and ideology figure in it very noticeably as derivatives of economic forces, and often get very little attention at all. Masses, for instance, tend to appear as categories resulting from the structural evolution of underdevelopment or dependent development; thus landlords are technically backward, domestic bourgeoisies are weak and comprador in outlook, wage-labour forces are small and highly differentiated, etc. Classes do not appear as the prime movers of history in this perspective, and a corresponding theoretical eclecticism is apparent ('élite' is often used alongside 'class', for instance). And UDT tends to be mechanical in the sense that processes tend to be presented as resulting from a 'logic' or mechanism, a system of vicious circles reinforcing each other (e.g. low productivity means low real incomes, which mean narrow markets, which mean high levels of protection for manufacturing, which mean high-cost production which means lower real incomes — etc., etc.). This results in accounts that appear curiously static; 'underdevelopment' can appear as inescapable (something countries are 'locked into') as the 'breakdown' of capitalism appeared in the Marxism of the Second International. Little attention is paid to the struggle against imperialism and the social and political consequences of underdevelopment; still less do most UD theorists concentrate on this struggle as the key to overcoming underdevelopment. A good example of the confusion and vulnerability of UDT on something obviously central is the debate, stimulated by Bill Warren, as to whether 'underdevelopment' precludes rapid and 'independent' capitalist industrialization at the periphery, and, if not, whether (as Cardoso implies) the 'dependent capitalist industrialization' through MNC branch plants is ultimately different from other kinds of capitalist industrialization; if not, are we not back with 'late development' and 'catching up'?

Imperialism

On the whole this concept appears in UDT only as an 'extra', often with a qualifier (such as 'economic imperialism', 'cultural imperialism', etc.) and often meaning no more than the mechanisms and institutions of penetration and control employed by international capital in the Third World. There is a certain ambivalence, to say the least, towards the Leninist conception of imperialism as a historical stage of capitalism, an ambivalence which is, I think, primarily political and only secondarily theoretical; i.e. in the Third World context the adoption of such a concept implies a break with the perspective of reform which much UDT has not really made. But avoiding the concept frequently means avoiding realities that are central to the situations with which UDT is meant to deal.

Units of analysis

Another area where the conceptual imprecision of UDT is unacceptable concerns the units of analysis which should be central to it — social formations, modes of production, world economy, etc. The positive merits of UDT turn on its insistence on the importance of transnational or trans-state relationships in explaining the course of events in underdevelopment in Third World countries. Yet concepts used to analyse these are often far from clear. For instance, Frank's well-known characterization of the whole Latin American economy as 'capitalist' from the sixteenth century seems to preclude a systematic analysis of the class forces there (over that time they can't all have been either capitalists, compradors or proletarians — and are they yet?); yet the problematic of the 'articulation' of different modes of production, which is implied by Laclau's equally well-known critique of Frank, presents its own difficulties.

What causes underdevelopment?

Curiously enough, it is not clear that UDT provides any explanation of why more capital did not get invested and accumulated in the Third World in the past, or of why it should not now take advantage of cheap labour and soak up the vast pools of unemployed people in the Third World today. This is connected, of course, with the failure to be clear about the question of exploitation; it shows again how UDT offers only a general interpretation of a current situation, a system of mechanisms broadly taken as 'given', rather than putting forward a theory of the 'laws' governing the historical appearance and subsequent evolution of those mechanisms. The list could be extended. Booth's remark that Frank's formulations 'sensitise us to certain features of economic change and at the same time anaesthetise us to other features' can fairly be applied, it seems, to UDT generally. But the comment does not go far enough. The trouble with UDT is not that it is a theory still at an early stage of development, a theory with various gaps needing to be filled up and ambiguities to be resolved (as I myself used to suppose). The real trouble is that concepts which primarily 'sensitize' do not belong to a coherent theoretical system, based on a consistent political standpoint and a consistent method of analysis; they are ideological rather than scientific. This is why UDT is 'stalled', and why its real gains — the detailed analyses of the institutions and structures of underdevelopment — are being appropriated more by the ideologists of international capital than by the workers and peasants of the Third World.

UDT as 'Marxified structuralism'

In order to understand this better, we must look again at the origins of UDT, taking advantage of the excellent studies by Foster-Carter, Girvan, O'Brien and Booth.[7] What these all confirm is that the main line of descent of UDT, almost wholly a New World phenomenon, was a series of revisions of orthodox (neoclassical) economic analysis, beginning with the critique of international trade theory which was the starting-point of the ECLA 'school'. This initial revision explained the failure of Latin American and Caribbean economies to develop by identifying structural obstacles that would have to be overcome by appropriate policies to change the structures — policies summed up in the phrase 'inward-orientated development', and meaning particularly industrialization through import substitution. This was followed by a further revision that explained the failure of this strategy in terms of a new set of structural barriers, often structures which were introduced in the process of inward-orientated industrialization and which tended to defeat the object of the original restructuring policy (e.g. the high import content of MNC branch plant industrialization); or structural obstacles at another level, which were revealed in the course of the initial restructuring effort (e.g. highly unequal income structures, urban-rural imbalances, etc.), structures which in turn often seemed to be aggravated by the new strategies. 'Underdevelopment' and 'dependency' then emerged as concepts embracing all these newly apparent structures and mechanisms. At the same time, as the fresh 'layers' of structural obstacles were identified they came more and more to be seen as part of a long-term process involving successive stages, each with its distinctive institutional mechanisms reinforcing or modifying economic and social structures, but always reproducing their fundamentally subordinate character, serving development elsewhere, minimizing its benefits and maximizing its costs locally. The mainstream of UDT can thus be seen as eventuating in radical structuralism, i.e. as a structuralist analysis of the obstacles to capitalist development in the Third World, in which progressively more and more of what were originally seen as means to structural change — international manufacturing companies, Third World governments and the interests they mostly

[7] See notes 4 and 6 above; and A. Foster Carter, 'Neo-Marxist Approaches to Underdevelopment', in E. De Kadt and G. Williams (eds), *Sociology and Development* (London: Tavistock, 1973), and N. Girvan, 'The Development of Dependency Economics in the Caribbean and Latin America', *Social and Economic Studies* 22/1, 1973, pp. 1–33.

represent, etc. — come to be seen as yet further structures that themselves need to be changed.

The central weakness of radical structuralism is obvious. Inverting Cleaver's dictum that 'if you're not part of the solution you're part of the problem', radical structuralism cannot escape the difficulty that the solution always turns out to be part of the problem. This is particularly clear in the case of Gunnar Myrdal, who is not a dependency or underdevelopment theorist but who certainly is a structuralist and a social-democrat, and whose *Challenge of World Poverty* (New York: Pantheon Books, 1970) identifies as main causes of poverty the very governments and international agencies to whom he also appeals for solutions. In the case of the earlier writings of Best, Girvan and their collaborators in the West Indies, it appears in the shape of a reform programme for an anti-imperialist government based on an alliance of progressive local capitalists with the popular masses, which was, in fact, just what the existing governments of the West Indies generally claimed to be.[8] Neither Myrdal nor the New World economists appear to suggest how the problem is to become the solution; their utopianism is quite clear.

But in the case of UDT the central weakness is partly concealed, thanks to the convergence of its mainstream with another stream — Marxism, and especially neo-Marxism of various kinds. In the case of Frank, there was the influence of Paul Baran, whose *Political Economy of Growth* appeared in 1957; and, as Frank himself says, the influence of the Cuban revolution. More generally there was the impact of de-Stalinization and polycentrism on the Latin American Communist parties, and the subsequent influence of Maoism added to the already complex pattern of Latin American Marxism, with its strong Trotskyist currents. But, whatever the diversity of channels, the Marxist stream certainly mingled with that of structuralism in the 1960s, to the point where O'Brien, for instance, could refer to them as 'different *traditions within* the theory of dependency'.[9]

The immediate effect of 'Marxifying' radical structuralism was to shift the implied solution to the problem of underdevelopment from economic nationalism to socialism, and to shift the means from utopian recommendations of radical structural reforms to revolutionary struggle. But this was an illusion. It is no less utopian to appeal to 'revolution' and 'socialism' to solve the problem as radical structuralism formulates it than to appeal to the existing Third World governments or the US Agency for International Development (USAID), since a structuralist analysis doesn't disclose the potential class forces on which a revolutionary struggle can be based or the contradictions that condition and are developed by the struggle; nor does it disclose a strategy or organizational forms of struggle, or — *a fortiori* — a

[8] See Girvan, ibid., and L. Best, 'Outline of a Model of Pure Plantation Economy', *Social and Economic Studies* 17/3, 1968, pp. 283–324.

[9] O'Brien, 'Critique of Latin American Theories', op. cit., p. 11, (my emphasis).

'socialist' solution, since a socialist solution must itself be disclosed by the interests and capacities of the revolutionary forces and their strategy, which have not been identified at all. In short, the 'Marxification' of radical structuralism, or 'left' UDT, does not rescue radical structuralism from its dilemma, for the basic reason that it has been built up through successive revisions of bourgeois economics, and, like a Russian doll, the final outer layer has essentially the same shape as the innermost one. It is an analysis whose central concepts and problematic are ultimately still those of bourgeois social science, especially bourgeois economics, in spite of the fact that it uses words that denote the concepts of historical materialism; i.e. it understands the relations of production and exploitation in the Third World in a way which still idealizes and mystifies them.

The converse of this, of course, is that the 'Marxism' of the most Marxist of UD theorists becomes utopian. This seems to me the main difficulty with Frank's political position, at least as it emerges from his writings. Emphasizing revolutionary class struggle for socialism as the only solution to underdevelopment, he rejects dogmatic formulae for this struggle and recognizes the great variety of patterns and stages of underdevelopment in the different countries of Latin America. But he does not himself offer much in the way of analysis which would make the reality and meaning of this 'solution' more apparent and concrete. The main thrust of his work is a historical-structural analysis of surplus extraction and transfer, and structural subordination of periphery economies through the mediation of subordinate (or 'lumpen') bourgeoisies, not an analysis of the structures of oppression and exploitation, i.e. the modes and relations of production, and the class struggles and class organizations developing within these. Kay's polemical comment cited at the beginning of this chapter would be unjust if applied to Frank without qualification, but there would still be an element of truth in it.

What is more, 'Marxist' or 'left' UDT is not merely utopian in its own way, but ultimately cannot escape being ideological. An example of this is my own idea that some of the limitations of UDT of which I had become somewhat aware in my work on Kenya were 'gaps' which could be 'plugged' by grafting on a treatment of what was missing, e.g. an analysis of classes. The result was distinctly lifeless, uninspired by any real concept of class struggle.[10] In order to arrive at an analysis of classes for the purpose of relating class struggle to the problem of development, I needed to break away from the problematic of structuralism itself. So long as I remained within it, I remained a prisoner of its ideological effects, including a concept of social class which is ultimately residual and passive. More significantly, I would now say that I think UDT itself is really ahistorical, in the same way that bourgeois development theory is, and for the same reason (i.e. its problematic is the same). It does have a time

[10] See Colin Leys, *Underdevelopment in Kenya* (Berkeley: University of California Press; London: James Currey, 1975), especially Chapter 6.

dimension, of course, which some versions of bourgeois development theory conspicuously lacked. But this is not the same thing as history. The mere extension of the model to allow for change, or to see the present as the end of a chain of events extending back into the past, is not to make it historical, i.e. does not explain underdevelopment by reference to social forces which have been identified as a result of a coherent and consistent method for sifting out from the complex flow of events the patterns that allow us to make sense of it, now and in the past. The stages familiar from UDT (such as mercantilism, liberalism, neo-mercantilism, etc.) are rather empirical periodizations with at most a good deal of descriptive and heuristic value, not the product of a systematic analysis of the interplay of political and social as well as economic forces which resulted in the geographical extension of the sway of capital, or of the new struggles generated by this process.

Most of the points above are made, implicitly where not explicitly, in Frank's 'answer to critics', where he also makes the illuminating observation that the concept of 'dependence' can best be understood as the product of a very specific historical period, namely the long post-war 'boom', or phase of rapid capital accumulation;[11] it refers, in fact, to that boom, as seen from Latin America. The end of that phase of the accumulation cycle, he remarks, shifts attention to its cyclical nature, and the concept of 'dependence' necessarily gives way to the concepts needed in order to understand not just one phase, but 'the process of capital accumulation in the world — with special reference to the participation of Asia, the Middle East, Africa and Latin America — from 1500 to the present', a task he has set for himself and which, he notes, is also being undertaken in different ways, by others, such as Laclau, Hinkelammert, Marini and Torres (and we should presumably add Amin, Wallerstein and Anderson).

In the meantime, the ideological character of UDT is best demonstrated by its vulnerability to what Kay calls 'conservative reabsorption' of the kind represented by the World Bank's adoption of a 'poverty-orientated' aid philosophy under McNamara's presidency, and the corresponding 're-orientation' of bilateral aid doctrines by the USA and other countries. What this comes down to is the adoption of a non-radical structuralism. This says, in effect, that not all solutions are part of the problem: the World Bank isn't, nor are MNCs (if Third World governments pool their bargaining know-how and co-ordinate their collective leverage), nor are all Third World governments, at least not monolithically. On the other hand: (i) radical structural change is unlikely; (ii) it usually turns out to be very unpleasant (to whom is not usually made clear) and unpredictable in its outcomes; and (iii) who are we (in the bourgeois democracies) to advocate such radical changes for others (etc., etc.), so let us pursue structural reform

[11] A.G. Frank, 'Dependence is Dead, Long Live Dependence and the Class Struggle: An Answer to Critics', *Latin American Perspectives* 1/1, 1974, pp. 87–106.

instead.[12] This position is, in my view, simply the expression of a political programme of large-scale international capital and the states and international agencies that represent it. The assimilation of structuralism is at best partial in such perspectives, but, even if it were more complete, it would still be what it is — a form of bourgeois ideology, whose character has been very adequately revealed by the critiques of Baran and Amin and the polemics of Frank.[13]

We must therefore go forward, not backward; not merely 'beyond the sociology of development' (the title of Ivor Oxaal, Tony Barnett and David Booth's book (London: Routledge and Kegan Paul, 1975)) but beyond the sociology of underdevelopment in two ways: not merely into a different kind of theory, but into a more clearly defined relation between theory and practice. A genuinely historical theory will allow us to analyse the process of combined and uneven development of capitalism on a world scale, as it has been experienced (or 'participated in' as Frank rightly puts it) in particular countries, and hence as it presents itself to any one of them now, in the form of a class struggle conducted in the framework of a particular inherited structure of productive relations, forms of exploitation and exchange relationships, and a particular structure of relations of political and ideological domination, internal and external, etc. A serious and systematic application of historical materialism does not mean the application of a ready-made formula taken from the texts of Marx and Engels, or Lenin or Trotsky or Mao Tse-tung, but the application of the method and conclusions to be drawn from such texts plus the practice out of which they emerged, and by which they have since been continually revised.

And this is why it is not merely a question of a different kind of theory (i.e. different from UDT or disguised radical structuralism) but also of a more clearly defined relation between theory and practice. The reason I am not too happy about Frank's conclusion that UDT should give way to the analysis of the process of capital accumulation from 1500 AD to the present is that it seems even more liable than UDT to be an academic exercise in the literal sense of the term, and hence liable to be still more vaguely related to political practice, and hence, finally, liable to take on an ideological character all over again. It is not that such an analysis is unnecessary; it is very necessary. But what is involved is hardly less than a history of the modern world; and any such product conceived of primarily as a product

[12] A study of McNamara's speeches in the early 1970s shows that a remarkably large proportion of the main theses of UDT were incorporated in them, although in deradicalized versions. For an official systematization, see the International Bank for Reconstruction and Development/Institute of Development Studies (IBRD/IDS) study, Hollis Chenery *et al.*, *Redistribution With Growth* (London: Oxford University Press/ILO, 1975), which is criticized, with replies by some of the authors, in C. Leys, 'The Politics of Redistribution With Growth', *IDS Bulletin* 7/2, 1975.

[13] See S. Amin, *Unequal Development* (Hassocks: Harvester Press, 1976); and A. Gunder Frank, 'Sociology of Underdevelopment and Underdevelopment of Sociology' in his *Latin America: Reform or Revolution* (Greenwich, Connecticut: Fawcett, 1968).

of scholarship (let alone individual scholarship) or 'social science' (to use Frank's term) rather than as a work of synthesizing the experience of the political practice of particular classes on a world scale, and hence proceeding by a process of provisional formulations based on the experience of specific struggles, and so necessarily consisting of a protracted and collective endeavour, seems to me likely to prove essentially ideological. Perhaps Frank himself should not be interpreted too literally when he seems to propose the task in these terms, but, without going into detail over the examples we already have (such as Wallerstein's and Anderson's work), I think it will be important to envisage the development of a historical context for future theory by way of successive approximations, arising out of studies of historical situations specific enough for the study to be related in a very explicit way to the contemporary class struggles within them.[14]

Some theoretical questions

In conclusion, and in the context of what I have just argued, I wish to raise three theoretical issues from among the many which have to be confronted: the question of exploitation, the question of modes of production and their 'articulation', and the question of the reasons for underdevelopment.

Exploitation

UDT generally holds or implies that the condition of the people in underdeveloped countries is the result of exploitation, and this certainly appeals to 'common sense' as one contemplates generally the record of colonialism in, say, Peru or the Congo. But Marxist economists such as Bettelheim and Kay rightly question 'common sense'. While there may be 'superexploitation' in the periphery, i.e. extraction of surplus value from workers through coercive sanctions in precapitalist modes of production, or in the course of primitive capital accumulation (forced labour, etc.), the direct exploitation by capital of wage-workers at the periphery is typically less than in the capitalist metropoles if the level of exploitation is the proportion of the value of what the workers produce that is not paid to them in wages. In the metropoles wage levels are much higher but productivity is higher still, thanks to the higher organic composition of capital. The problem of the underdeveloped countries is, then, Kay argues, that the workers are exploited without being exploited enough — in the sense that they are subject to severe downward pressure on wages, to extract the maximum absolute surplus value, but do not benefit from the higher real wages that they could be paid if capitalists were forced to try to increase relative surplus value by more investment leading to higher productivity. In fact, Kay argues, the

[14] I. Wallerstein, *The Modern World System* (New York: Academic Press, 1976); P. Anderson, *Passages from Antiquity to Feudalism* and *Lineages of the Absolutist State* (both London: New Left Books, 1975).

value of wages may be lower in the metropoles than at the periphery, even though wages are much higher there in terms of purchasing power — meaning that the labour time necessary to produce the goods their wages buy is probably smaller than the time taken to produce the relatively few goods the Third World workers can buy with *their* wages.[15]

Against this, Emmanuel argues that it is artificial to treat the metropoles and the periphery countries in isolation from each other in this way.[16] If a pair of shoes can be produced anywhere in the world in ten minutes of labour time, that is the (world) value of a pair of shoes and the proper measure of the value of real wages everywhere; the tendency of this interpretation is to make the real wages of Third World workers seem much lower as a proportion of what they produce, and hence to make their level of exploitation higher. By adopting this general standpoint one can also argue that, merely because sufficient capital has not been invested to make workers as productive as workers employed in similar work elsewhere, it does not follow that they are less exploited. If capitalists choose in certain areas not to extract the maximum surplus value from the workers that is possible using the best available equipment and organization, that is because in those areas political, social and economic conditions permit them to pay such low wages as to make the additional investment unnecessary. This line of argument has, it seems to me, considerable force, and corresponds to our 'common-sense' feeling that low-paid workers in undercapitalized operations are more exploited than higher paid workers in more efficient plants in the same line of business.

At all events, it is not hard to see that these arguments all have a rather abstract quality and that what is at issue is not so much a 'correct' theorization of the locus and relative levels of exploitation as between First and Third World countries and workers, as certain practical-political implications that have been supposed to follow from one position rather than another. Thus Bettelheim, for instance, was really anxious to show that the exploitation of workers in France or the USA was the source of the economic power behind imperialism and hence of the oppression (and hence 'superexploitation') of workers in the Third World, and to discourage the idea that Third World workers had a common cause with the bourgeoisies and petty bourgeoisies of their own countries against both foreign capital and foreign workers (who were considered by Emmanuel to be sharing in the exploitation of Third World workers through the operation of the terms of trade).

Perhaps by now we are free to dislodge the discussion from such an abstract and universal plane. Is it really necessary to hold that all workers in all underdeveloped countries are more or less exploited than all workers in all the metropolitan countries? Is it not clear that the technical concept of exploitation, formulated by Marx in the context of the

[15] Kay, *Development and Underdevelopment*, op. cit., pp. 50–54 and 115–16.
[16] Emmanuel, *Unequal Exchange*, op. cit., pp. 382–83.

competitive capitalism of a single social formation, cannot be stretched to cover unambiguously all the complex relationships between workers and capitalists and workers and workers throughout a world capitalist economy? So, for instance, when an MNC moves its television-manufacturing operation from the USA to Taiwan it does so precisely to reduce the overall level of real wages it pays and so increase absolute surplus value. The Taiwanese workers are evidently more exploited than the US workers whom they replace, using identical equipment. They are also likely to have higher real incomes than Taiwanese workers in plants with much lower organic composition of capital and who as a result produce less per hour of labour. Which of these two groups of workers is considered more exploited depends both on relative real wages and relative productivity, but also on the position one takes on the general theoretical issue discussed above. Furthermore, both are directly exploited in a way that the Taiwanese peasant producer selling his/her crop to realize a still lower real income is not. All three groups of workers are in differing ways victims of an international and national structure of social relations and political domination which condemns them to (among other things) a level of real incomes which the development of the forces of production on a worldwide basis has long since rendered unnecessary. The practical political implications of these theoretical distinctions are not unimportant, affecting workers' and peasants' solidarity and the possibilities of international class co-operation too; but they are only part of the ensemble of relations which also affect solidarity, consciousness, organization, etc. In short the general question about exploitation in the Third World needs to be dissolved into a series of more particular questions about the forms, degrees and effects of different kinds of exploitation in different national and international contexts, and the way these relate to forms, degrees and effects of domination and oppression and the struggles waged against both.

Units of analysis: modes of production and their 'articulation'

As soon as we move beyond the broad perspective of UDT according to which the condition of the 'periphery' is accepted as simply being, as a matter of fact, complementary to the process of capital accumulation at the centre, two questions arise: (i) how; and (ii) why? Frank did not really propose an answer to the second question (nor did Baran). Frank's answer to the first question was that, in Latin America, it was done by means of changing forms of monopoly through a hierarchy of centres of surplus expropriation/appropriation (extraction and transfer), the top of which lay in the capitalist metropoles; and that the bottom of the hierarchy consisted of institutions established for this purpose, hence capitalist institutions, not feudal institutions (the latifundium was 'born as a commercial enterprise').[17] The conditions of servitude on the latifundium were only 'feudal-seeming',

[17] A.G. Frank, *Capitalism and Underdevelopment in Latin America* (New York: Monthly Review Press, 1967).

for the latifundium was not an institution 'beyond the reach of capitalist development'. The practical implication of this view seemed to be that the task of the revolution was not one of first overthrowing feudalism in Latin America, and only then going on to overthrow capitalism; and hence it was not a question of the proletariat and peasantry allying with the national bourgeoisie against feudal landlords; on the contrary, the feudal-seeming landowners were really capitalist and both they and the so-called 'national' bourgeoisie were playing and had always played a subordinate role in alliance with imperialism, so that the struggle of the workers and peasants must be waged against all of them, and must be a struggle to move from capitalism to socialism.

Laclau in his well known critique pointed out that: (i) this conceptualization includes in 'capitalism' all productive processes, and the social formations dominated by them, which are in a significant exchange relationship with capitalist social formations (so that, for example, Russia in 1850 would presumably have to be described as 'capitalist'); and (ii) capital, in the form of merchant capital, was being accumulated long before capitalism — i.e. social formations dominated by the capitalist mode of production — was established; in particular, long-distance trade and commodity markets existed long before capitalism.[18] The colonization of Latin America was in fact a feudal colonization, the response of Iberian feudalism to a general feudal economic crisis which could be relieved (although not resolved) by extending the area of land under servile cultivation through overseas expansion. The latifundium was not a 'feudal-seeming' but really a feudal institution. Subsequently, with the establishment of capitalism as a dominant mode of production in parts of Western Europe, there arose a world capitalist system, i.e. primarily a system of (unequal) exchange linking social formations dominated by precapitalist modes of production, including those of Latin America, to those dominated by the capitalist mode of production. The effect was to consolidate the precapitalist modes of production, not to destroy them (for reasons touched on below). This means that the first revolutionary task in such situations is to overthrow the precapitalist relations of production and the corresponding superstructures. Specifically, it means that in some parts of the world the revolution has to accomplish some of the tasks historically undertaken by the bourgeois-democratic revolution in Europe before the task of building socialism can be tackled — liberating servile labour, democratizing national and local institutions, abolishing all kinds of privileges and taboos, destroying archaic forms of ideological domination, etc. It follows that any call for 'socialist revolution' which ignores this, and the strategy, organization, programme, alliances, etc. which it implies, will be utopian. At any rate, Laclau argues, such vital issues cannot even be posed in terms of the formulations used by Frank.

This has been endorsed in various forms by writers such as Bettelheim

[18] E. Laclau, 'Feudalism and Capitalism in Latin America', *New Left Review* 67, 1971, pp. 19–38.

and Rey, who can be lumped together as theorists of the problematic of the 'articulation' of different modes of production.[19] According to this, the way in which the capitalist mode of production is 'articulated' with precapitalist modes of production that continue to exist in the era of capitalism constitutes, in Rey's terms, the 'field of contradictions' of the class struggles in the social formations where this articulation occurs (above all, in underdeveloped areas). This field is determined by capitalism's need to 'conserve' the precapitalist modes of production at the same time as it 'dissolves' them (i.e. it transforms them from modes of production subject to their own 'laws of motion' to modes of production whose motion is primarily determined by the laws operating in the capitalist mode of production, and hence also progressively transforms the content, and eventually the form, of their relations of production).

Reacting against this in turn, Banaji has cogently argued that writers of this school, including Laclau, reduce the meaning of 'mode of production' to that of 'mode of exploitation', or 'form of the labour process', a 'simple category' of analysis, like an ideal type, formed by abstracting a few empirical features from a given situation.[20] Thus, a 'feudal mode of production' is held to exist whenever labour is serf-like; but, Banaji says, one slave plantation does not make a slave mode of production, nor yet do quite a number. What constitutes a slave mode of production in the historical sense is a much wider ensemble of relations, including class, state and ideological relations, which together are governed by determinate 'laws of motion' which at the level of any given enterprise (in this case, a plantation) are revealed in the logic of its operations, the economic motivation of its owners and managers (production for social consumption, not for profit), the specific rationality of its production, consumption, savings and investment decisions, the degree of exploitation of its labour force, etc. On this basis, the plantations of the West Indies were capitalist, notwithstanding the reality of the slavery on which they rested;[21] the twentieth-century plantations established in Peru by large US and British firms using forms of servile labour are capitalist, not feudal; and the peasants of Kenya who produce leaf for capitalist tea companies at a closely controlled price are a kind of wage-labourers, not participants in a 'peasant mode of production'. In other words, Banaji agrees with Laclau that the Latin American latifundium was a feudal institution but disagrees with him in thinking that the persistence of its feudal form necessarily implies the continuance of its feudal content once the 'laws of motion' of feudalism, to which it was once subject, have been

[19] See C. Bettelheim in Appendix I to Emmanuel, *Unequal Exchange*, op. cit., pp. 293–99; and P.-P. Rey, *Les Alliances de classes* (Paris: Maspero, 1973).

[20] J. Banaji, 'Modes of Production in a Materialist Conception of History', *Capital and Class* 6, 1977, pp. 1–44.

[21] An interpretation which seems to have the explicit endorsement of Marx: 'The fact that we now not only call the plantation owners in America capitalists, but that they *are* capitalists, is based on their existence as anomalies within a world market based on free labour' (*Grundrisse*, (London: Penguin, 1973), p. 513).

displaced by the 'laws of motion' of the capitalist mode of production, as increasingly occurred from the seventeenth century to the nineteenth. The political implication of this seems to be that the nature of the revolutionary task is different in cases where a precapitalist mode of production has really survived from those where its forms may have survived and even become more 'highly developed' (as Banaji considers happened in Europe to feudal production precisely when it began to be subordinated to capitalism), but where its laws of motion have ceased to be at all autonomous. These general formulations sometimes have a rather abstract and even scholastic quality, but it seems to me they must be taken seriously. The alternative is only new forms of utopianism. At the same time, I think both Laclau's and Banaji's presentation of the matter suffers from a neglect of its political aspect, i.e. of class struggle and class domination (including state power). This does not mean that this aspect is 'given' and can be known independently of the determination of the mode(s) of production, forms of labour process, etc.; but just as Banaji thinks that modes of production cannot be determined merely by reference to the mode of exploitation but must be 'living' ensembles of relationships with a definite 'logic' and historical tendencies, so I do not think they can be determined independently of the determination of the classes, forms of state, etc., which are based on them and which also determine their development. (The same point was made in another context by Bettelheim in his debate with Sweezy.)[22] I suspect, in fact, that the apparent abstractness (and some of the difficulty) of the problem of deciding whether a precapitalist mode of production still 'exists' in a particular situation, or just how far and in what ways it is being transformed into a mere subordinate form of the capitalist mode, disappears when it is firmly linked to the problem of exploring the past and present course of class formation and class struggles.

Why does capitalism 'underdevelop' some countries?

Laclau pointed out that UDT really sidesteps this issue; Frank, in particular, sees a single contradiction between the bourgeoisie in the metropole and the masses in the periphery from the sixteenth century till today. But if the initial dependency was one within feudalism the reasons for it cannot be identical with those that explain capitalist dependency. Laclau holds that the initial feudal dependence turned on unequal exchange between Latin America and Europe, made possible by holding down price levels in Latin America by an extremely ruthless use of servile forms of labour, thereby 'reducing the economic surplus of the peripheral countries and fixing their relations of production in an archaic mould of extra-economic coercion, which retarded a process of social differentiation and diminished the size of their internal markets'.[23] Subsequent capitalist underdevelopment (i.e. once

[22] See P. Sweezy and C. Bettelheim, *On the Transition to Socialism* (New York: Monthly Review Press, 1971), pp. 15–17.

[23] Laclau, 'Feudalism and Capitalism', op. cit., p. 35.

the capitalist mode of production had become dominant in Europe in the nineteenth century) was due to a different central cause, the decline in the rate of profit caused by the increasing organic composition of capital. The maintenance of precapitalist modes of production in the colonies and semi-colonies, he hypothesizes, maintained the average rate of profit by offering opportunities for high rates of return due to the local organic composition of capital and the use of extra-economic coercion of labour in those modes. This hypothesis refers to the competitive phase of capitalism and explains capitalist underdevelopment in that stage only.

Kay, as I understand him, takes a different view. Labour at the periphery may have been (and may still be) subject to extra-economic coercion and also may be paid a much lower real wage than the 'free labour' of the metropoles but it does not follow from this that profit rates are higher; this depends on the relative rates of surplus value and ratios of constant to variable capital at the centre and at the periphery. Part of what seems to be at issue here is an empirical question, which Laclau poses: during the nineteenth century, was the growth in the organic composition of capital in fact more rapid than the growth in the productivity of labour in the metropoles? If it was, the profit rate would have tended to fall and the maintenance of precapitalist modes of production at the periphery would have tended to offset this for the reasons Laclau suggests. If it was not, the rate of profit would have tended to rise, not fall, and the only reason for capital to be invested in the periphery at all would be to secure particular commodities needed for production in the metropole (cotton, minerals, etc.) and cheap foodstuffs (reducing the value of the real wages paid in the metropole). Working on the latter assumption, Kay suggests that the explanation of capitalist underdevelopment is really historical and structural, and is to be sought in the special characteristics of merchant capital. Merchant capital, he argues, deriving its profits entirely from unequal exchange and from maximizing the flow of transactions, destroyed the pre-existing social frameworks in the periphery (Kay evidently has Africa particularly in mind), and actually created conditions inimical to the investment of industrial capital, which, once it gained supremacy over merchant capital in the metropoles, sought quite limited ends in the periphery (cheap raw materials and food supplies), using merchant capital as its agent. Playing this role of agent, merchant capital was forced to maintain the forms of precapitalist modes of production and social relations of production, in order to maximize the production of commodities for metropolitan markets and maintain conditions for regular and reliable trading.[24]

Although there seems to be some difficulty with both these views, they have the great merit of being genuine theories, and I suspect that if they are understood as theories applicable to more limited places — i.e. Latin America and Africa respectively — and to specific phases of metropolitan

[24] Kay, *Development and Underdevelopment*, op. cit., pp. 105–7.

development, they will prove capable of contributing to the sort of general theory that is needed.

Concluding remarks: theory and practice

Consideration of these theoretical questions seems to suggest once more that a large part of the difficulty of resolving many of them is due to the same cause that is ultimately responsible for the ideological nature of UDT itself — a particular form of political practice. In the case of UDT, that practice, it is safe to say, was primarily that of planners in the state apparatuses of Third World countries, of their advisers and mediators (with international capital) in regional economic policy bureaux, such as ECLA, and of the personnel of the 'aid and trade' network, from the UN Conference on Trade and Development (UNCTAD) to the International Bank for Reconstruction and Development/International Monetary Fund (IBRD/IMF). The UD 'perspective' really expressed the 'unhappy consciousness' of progressive intellectuals (including academics) within or close to this set of institutions, on or just over the dividing line between reluctant acceptance and impatient rejection of the structure, but not starting out from or conditioned by involvement in the entirely different practice of a radical working-class or peasant party. The ambiguities of UDT reflect the ambiguities of their position.

The work of the critics of UDT, on the other hand, while it may be much less ambiguous, does show a tendency to theoreticism, abstractness and generality, reflecting perhaps the primarily academic practice of most of those concerned. There is, certainly, no uniquely certified correct connection between theory and the political practice of the exploited and oppressed classes, and the problem of making any such connection is aggravated by the difficulty, in many parts of the Third World, of identifying effective current forms of popular politics and establishing links with them. None the less, it looks as if it will be very important to resolve this question as a condition of useful theoretical advance.

3

Samuel Huntington & the End of Classical Modernization Theory

In the conceptual fragmentation and conflict that has overtaken development studies, the leading place has often seemed to be held by the 'dependency' or 'underdevelopment' theorists. This school first became widely known in the West toward the end of the 1960s, particularly through the polemical attack led by Andre Gunder Frank on US-dominated 'modernization' theory. Since then the *dependencistas* have been severely criticized in their turn by Marxists, but there would appear to have been no serious reply from the orthodox right.

This appearance is deceptive, however, especially in the USA and perhaps above all among political scientists in the US sphere of influence — which is still a considerable one. It is true that the ideologists of developmentalism had no answers to the devastating critique of Andre Gunder Frank, Suzanne Bodenheimer and others.[1] However, their 'paradigm-enforcing' powers were immense. For a long time the *dependencistas*' attack was simply ignored.[2] Instead of a scientific debate, there was a retreat from confrontation with anti-imperialist theory, while efforts were made to consolidate a major transformation in mainstream development theory that was already under way when the onslaught by Frank and others occurred.

* First published in Hamza Alavi and Teodor Shanin (eds), *Introduction to the Sociology of 'Developing Societies'* (New York: Monthly Review Press, 1982), pp. 332–49.

[1] Andre Gunder Frank, 'The Sociology of Underdevelopment and the Underdevelopment of Sociology', *Catalyst*, Summer 1967, also in Frank, *Latin America: Underdevelopment or Revolution?* (New York: Monthly Review Press, 1969), pp. 21–94; Suzanne-J. Bodenheimer, 'The Ideology of Developmentalism: American Political Science's Paradigm-Surrogate for Latin American Studies', *Berkeley Journal of Sociology* 15, 1970, pp. 95–137; Gail Omvedt, 'Modernization Theories', in A.R. Desai (ed.), *Essays on Modernization of Underdeveloped Countries*, vol. I (Bombay: 1971); Henry Bernstein, 'Modernization Theory and the Sociological Study of Development', *Journal of Development Studies* 7, 1971, pp. 141–60.

[2] As far as I can tell no review of Frank's first three books (all published in the USA) has ever appeared in the *American Political Science Review*, *Journal of Politics*, *World Politics*, *Comparative Politics* or *Economic Development and Cultural Change* — the last of these being the journal whose school of thought Frank explicitly attacked in his celebrated 1967 critique of the modernization perspective.

This was a shift to a concern with the maintenance of order. The revolution in Cuba, followed by the accumulating humiliation of US policy in Vietnam and the revolutionary instability and increasing reaction in Latin America and South Asia, undermined the original optimistic assumption of orthodox development theory, that the process of development involved drawing the populations of the Third World out of their traditional isolation into a modern social system that would be participative, pluralistic and democratic. A growing number of US social scientists began to recognize that this was not in fact the destiny of Third World countries developing in the framework of US hegemony. Increasingly, their thinking turned from a concern with the processes of 'modernization' (the processes of transformation into advanced, capitalist and hence pluralist and democratic societies) to a concern with the maintenance of social control.[3]

Yet a complete break with the theoretical apparatus inherited from the optimistic phase of development ideology was neither practicable nor desirable. The need was to transform that apparatus — whose central animating concept was modernization — in such a way as to incorporate the new concern with maintaining order and, wherever possible, at the same time 'co-opt' the more persuasive themes of dependency theory and even Marxism itself. How this task has been carried out, in various branches of theory, lies beyond the scope of this essay. What is attempted here is to bring out the general nature of the ideological reconstruction involved by examining the work of one political scientist who, more than any other, led the way in initiating it.

'Modernization'

In order to explain Samuel Huntington's contribution, it is first necessary to sketch the main elements of 'modernization' theory itself.

The distinction between 'traditional' and 'modern' societies was derived from Max Weber via Talcott Parsons. A society was called 'traditional' in which most relationships were 'particularistic' rather than 'universalistic' (e.g. based on ties to particular people, such as kin, rather than on general criteria designating whole classes of persons); in which birth ('ascription') rather than 'achievement' was the general ground for holding a job or an office; in which feelings rather than objectivity governed relationships of all sorts (the distinction between 'affectivity' and 'neutrality'); and in which roles were not clearly separated — for instance, the royal household was also the state apparatus ('role diffuseness' vs. 'role specificity'). A society in

[3] For this context see the excellent early commentary by D.C. O'Brien, 'Modernization, Order, and the Erosion of a Democratic Ideal', *Journal of Development Studies* 7, 1971, pp. 141–60; also M. Kesselman, 'Order or Movement? The Literature of Political Development as Ideology', *World Politics* 26, 1973, pp. 139–54.

which the opposite of all these was true was 'modern'.[4] Other features generally seen as characteristic of traditional societies included things like a low level of division of labour, dependence on agriculture, low rates of growth of production, largely local networks of exchange and restricted administrative competence. Again, modern societies displayed the opposite features.

'Modernization', then, referred to the process of transition from traditional to modern principles of social organization, and this process was what was currently occurring in Asia, Africa and Latin America.

From today's vantage point it is easy to see that the traditional/modern distinction was at bottom simply a pair of very arbitrary abstractions from the contemporary situations of the periphery and the centre (respectively) of the world capitalist system; and that the concept of modernization was no more than a proposal to consider the former as a once-universal original state and the latter as a universal end state, with the so-called modernization process serving as a (quite fictitious) surrogate for the historical processes that have actually taken place, or are taking place, at both the periphery and the centre. However, those political scientists who were unaware of the ideological nature of these formulations saw in them the basis for an attempt to conceptualize 'political development' as 'one element of the modernization syndrome', i.e., the modernization process itself at the level of politics.[5] This gave rise to a multiplicity of proposals: political development, it was suggested, involved democratization, political 'mobilization', the 'building' of nation states, administrative and legal development, secularization, equality, 'sub-system autonomy', etc.

According to Huntington himself, the starting-point for his own theorizing was a rejection of the concept of 'political development' because it indeed had no clear meaning.[6] In its place he proposed to discuss merely political change. If this had been true, it would have implied at least an attempt to break with 'modernization' theory. But it was not true. In his main book, *Political Order in Changing Societies* (Yale University Press, 1968), a central place is occupied by the concept of 'political decay', a condition signalized by unrest, violence, corruption and coups, and one which retains all the normative and teleological content of the concept of 'political development', only expressed as its opposite.[7] And this continuity, this preservation of the essential concepts of modernization theory but in forms

[4] T. Parsons, *The Social System* (London: Tavistock Publications, 1952), pp. 58–67; also Frank, 'Sociology of Underdevelopment', op. cit., pp. 24–25.

[5] Huntington, 'The Change to Change: Modernization, Development, and Politics', *Comparative Politics* 3, 1971, pp. 283–322.

[6] Ibid., p. 301.

[7] An earlier statement of the central argument of the book was entitled 'Political Development and Political Decay', in *World Politics* 17, 1965 pp. 386–430. The continued presence of the concept of 'political development' in the 1968 book is also marked by the fact that the term itself often recurs (see, for example, pp. 7, 192, 398). Huntington's later claim to have dropped it was quite spurious.

adapted to the new concern with order, is, as we shall see, a fundamental aspect of Huntington's thought. However, in three respects Huntington did make a major break.

First, he really did switch attention to a kind of political change — revolution — which was the central issue of contemporary Third World politics. Previous writers had, remarkably, largely failed to do this. The concepts with which they proposed to analyse Third World politics were thinly disguised abstractions from the pluralist interpretation of US politics, according to which politics consists of incremental adjustments to the evolving balance between competing interests, mediated by competing electoral machines which 'aggregate' the interests of 'groups' drawn from all 'strata' of society. In Huntington's conception of politics, in contrast, the central place was given to potentially revolutionary struggles for state power by radically discontented social forces — workers, students, urban 'middle classes', peasants, etc. Huntington's view of Third World politics seemed above all realistic.

Second, Huntington's work seemed genuinely theoretical. It was very difficult to extract from the previous literature on 'political development' anything that really proposed explanations of large-scale, important political phenomena. There were more or less elaborate classificatory systems or typologies, but very little usable theory. Huntington, in contrast, offered to explain, among other things, why pressures for change assumed a reformist or revolutionary character, why reforms or revolutions succeeded or failed, and why, in the absence of either reform or revolution, Third World countries tended to pass under personal, arbitrary and authoritarian rule ('praetorianism').

These two changes have a lot to do with the academic and practical influence of Huntington's work. Attention has been diverted from them, however, by the third departure that Huntington made from previous orthodoxy: his open and abrasive commitment to the maintenance of order as the supreme political value. As Donal O'Brien has shown, Huntington's views on this issue were not very different from those of numerous other leading US political scientists of the later 1960s, but they were more 'stringently and unambiguously expressed'.[8] Throughout his work there runs a strong current of dislike for the confusing, disturbing and contradictory aspirations of the masses, and an admiration for any 'élite,' bureaucracy or 'leadership' capable of containing, channelling and if necessary suppressing them. We shall return below to the significance of the particular form taken by Huntington's obsession with authority. For the moment, however, the important point is that it was in spite of this, as much as because of it, that Huntington's work achieved the influence it did. Refreshing as it might be to many members of the beleaguered US professoriate in 1968, the open fear of and distaste for the masses that

[8] O'Brien, 'Modernization, Order', op. cit., p. 368.

Huntington expressed was hardly a viable posture for most of them.[9]

Huntington's argument

Conventional wisdom among both social scientists and policy-makers in the USA, Huntington noted, held that the poorer people are, the more prone they will be to use violence to remedy their situation (the more 'unstable' politics will be). Accordingly, US policies toward the Third World had, at least in theory, been directed toward securing economic growth in the belief that this, coupled with reforms made possible by growing output, would lead toward political 'stability'. But, according to Huntington, the statistical evidence shows that there is more stability in countries with the lowest per capita incomes than in countries somewhat less poor. Huntington's interpretation of this is that, as 'modernization' (defined as 'social mobilization plus economic development') occurs, more and more people become politically active, or become active in ways that impinge more and more on the central government ('political participation'), because increased social mobility leads to raised expectations which at an early stage of economic growth cannot be met. Economic growth also leads at first to heightened inequality, which also prompts greater 'political participation'. Unless political institutions are capable of handling this expanded 'participation', it will assume destabilizing forms (demonstrations, strikes, riots or even armed struggle) and/or lead to corruption (equally liable to give rise to instability in the long run).

What determines whether the increased 'level of participation' produces these destabilizing consequences or not is the degree of 'institutionalization' of the 'polity'. Strongly institutionalized polities socialize (i.e. formally and informally induct) the newly participant citizens into the channels and norms of political action that are prescribed by the existing structures. This leads to 'civic' politics. 'Civic' politics may or may not take a democratic form: 'constitutional democracies and communist dictatorships are both participant polities'.[10] Participation, in other words, need not (and

[9] The open advocacy of counter-revolutionary positions by US academics following Huntington's example in the early 1970s, after the successive defeats for US power in the Third World, deserves to be remembered in the hour of the 'triumph of liberal democracy'; see, e.g., Nelson Kasfir, 'Departicipation and Political Development in Black African Politics', *Studies in Comparative International Development* 9, 1974, pp. 3–25, and Norman H. Keen, 'Building Authority: A Return to Fundamentals', *World Politics* 26, 1973–74, pp. 331–52. Henry A. Landsberger and Tim McDaniel ('Hypermobilization in Chile, 1970–1973', *World Politics* 28, 1975–76, pp. 502–41) concluded that: 'Historically, mobilization seems to have been more of a hindrance, i.e., a "bad thing", than a help to radical governments, at least after the old centres of power have been weakened with its help. The real problem is then whether "the masses" can be controlled. If the government has enough coercive power to do that...it may not need to mobilize the masses in the first place.'

[10] S. Huntington, *Political Order in Changing Societies* (New Haven: Yale University Press, 1968), p. 89.

Huntington thinks should not) mean popular control of government, but rather governmental control of the people through their 'involvement' in the organs of the 'polity' that make this possible. Weakly institutionalized polities, on the other hand, are easily overwhelmed by new 'groups', which enter politics ('participate') on their own terms, giving rise to 'praetorianism', in which 'the wealthy bribe, students riot, workers strike, mobs demonstrate, and the military coup [sic]'.[11]

Praetorianism — of which military rule is the usual form —may easily degenerate into chaos. In such a situation, revolution may become the only possible form of lasting political change — because only a social movement capable of making a revolution will be capable of establishing a new, durable political order. Revolution, then, is for Huntington a form of political change that results from a severe discrepancy between the scale of the forces newly participating in politics and the capacity of political institutions to assimilate or contain them.

Revolutions are not, according to Huntington, primarily about economic issues: 'A revolution is a rapid, fundamental and violent domestic change in the dominant values and myths of a society, in its political institutions, social structure, leadership and government activity and policies.'[12] 'Ascending or aspiring groups and rigid or inflexible institutions are the stuff of which revolutions are made.'[13] The conditions leading to revolution are the simultaneous 'alienation' from the existing political order of the urban middle class and the peasantry, and their ability to collaborate, usually on a nationalist programme. These conditions are rarely adequately met, however, because of the tendency of the urban middle class to become more conservative as it grows larger, and because of the basic divergence of aims between it and the peasantry. The end-point of revolution is 'the creation and institutionalization of a new political order'; 'the measure of how successful a revolution is is the authority and stability of the institutions to which it gives birth'.[14] In this respect the communists have proved the most successful revolutionaries, because Leninism equipped them with an effective theory of political organization and so made possible a new political order in which mass participation is combined with 'a government really able to govern'.[15]

The important question about any reform, for Huntington, is not its merits *per se* but simply whether it averts revolution or acts as a catalyst for it. The Alliance for Progress in Latin America was based on the idea that reforms would avert revolution. Huntington, however, thinks that in some circumstances reforms may have the opposite effect. In particular, reforms that respond to the demands of the urban intelligentsia make revolution

[11] Ibid., p. 196.
[12] Ibid., p. 264.
[13] Ibid., p. 275.
[14] Ibid., p. 266.
[15] Ibid., p. 342.

more likely because the intelligentsia are revolutionary from 'psychological insecurity, personal alienation and guilt, and an overriding need for a secure sense of identity':[16]

> Programs catering to the demands of the radical middle class only increase the strength and radicalism of that class. They are unlikely to reduce its revolutionary proclivities. For the government interested in the maintenance of political stability, the appropriate response to middle-class radicalism is repression, not reform.[17]

In contrast, reforms catering to peasant demands can avert revolution, because

> no social group is more conservative than a landowning peasantry and none is more revolutionary than a peasantry which owns too little land or pays too high a rental. The stability of government in modernizing countries is thus, in some measure, dependent upon its ability to promote reform in the countryside.[18]

Such, in the barest outline, is Huntington's central argument. Let us now examine it more closely.

A critique

To grasp the significance of Huntington's work we have to analyse it on two levels: first, from the standpoint of bourgeois social science and, second, from that of historical materialism.

Huntington's work as social science

AS SOCIAL SCIENCE THEORY

One commentator, reviewing Huntington's argument in the mid-1970s, called it 'an elegant, simple, lucid and promising theory'.[19] In fact, on closer inspection it proved to be vague, tautologous and circular. Its two central concepts, 'participation' and 'institutionalization', were nowhere adequately defined. 'Participation' included, apparently, any kind of conscious relationship whatever with the polity, no matter how ritualistic or empty, since, as we have seen, Huntington declared that Soviet citizens had 'a high degree of popular participation in public affairs'.[20] Conversely, 'institutionalization' was defined as 'the process by which organizations and procedures acquire value and stability', but this process was nowhere specified either. Instead,

[16] Ibid., p. 371.

[17] Ibid., p. 373. Those interested in the principle of value-freedom in social science should note that Huntington is careful only to draw this conclusion for governments interested in the maintenance of political stability. By implication, those interested in their own overthrow could disregard it.

[18] Ibid., p. 375.

[19] Gabriel Ben-Dor, 'Institutionalization and Political Development: A Conceptual and Theoretical Analysis', *Comparative Studies in Society and History* 17, 1975, p. 310.

[20] Huntington, *Political Order*, op. cit., p. 1.

Huntington put forward four 'measures', or tests, of the 'level of institutionalization of any political system': the adaptability, complexity, autonomy and coherence of its organizations and procedures. Adaptability, Huntington suggested, could be measured by the age of an organization or procedure: 'The probability that an organization which is one hundred years old will survive one additional year... is perhaps one hundred times greater than the probability that an organization one year old will survive one additional year.'[21] A moment's thought shows that this is absurd. Presumably recognizing this, Huntington added that another measure is the ability of an organization or procedure to survive challenges to its existence. Thus tsardom, passing the age test with flying colours, failed the 'challenge' test in 1917. But the challenge test, in this case, was a revolution — i.e. an 'explosion of participation.'[22] So the theory seems to state that revolutions occur when the institutionalization of the 'polity' is insufficient to handle the scale of participation, and the evidence that it is insufficient is that revolutions occur — i.e. it is circular. Huntington, however, suggests yet a third measure of adaptability, the ability of an institution (organization or procedure) to abandon challenged functions and replace them with others and so survive. But this reasoning too is circular. The evidence that an institution is adaptable is that it changes functions and survives: the evidence that it is not is that it fails to change functions and comes to an end. Tsardom and the French monarchy failed, Huntington says, while the British monarchy succeeded, in finding something else to do but practise absolutism.[23] The evidence of failure is once again the success of revolution, with the same circular consequences.

The other measures of institutionalization (complexity, autonomy and coherence) suffer from exactly the same difficulty. Thus the measure of 'autonomy' is whether social forces 'outside' the political system succeed in forcing themselves 'into' it or not; the measure of coherence is whether the institutions of the polity rest on a consensus sufficiently strong to preserve them; and so on.

In short, the thesis that political change is a product of the relation between participation and institutionalization turns out to be spurious because 'institutionalization' is defined in terms of the response of institutions to 'participation' (whatever that means). It is not possible to catalogue here the multiple other shortcomings of Huntington's argument considered as theory, nor is it really necessary: they have been carefully pointed out by others, although in muted, not to say somewhat deferential, terms.[24]

[21] Ibid., pp. 13–14.

[22] Ibid., p. 266.

[23] Ibid., pp. 15–17.

[24] See Ben-Dor, 'Institutionalization and Political Development', op. cit.; Dean C. Tipps, 'Modernization Theory and the Comparative Study of Societies: A Critical Perspective', *Comparative Studies in Society and History* 15, 1973, pp. 199–226; Kesselman, 'Order or Movement?', op. cit.; and Dexter W. Lethinen, 'Modernization, Political Development, and Stability', *Stanford Journal of International Studies* 9, 1974, pp. 219–45.

EVIDENCE

Part of the reason for the failure to push home the criticism of Huntington's work as unscientific may well have been the massive quantity of facts deployed in it. 'Dazzlingly comprehensive'... 'his empirical knowledge seems encyclopedic' — most readers, like these reviewers, are intimidated by a great show of scholarship, and Huntington spared no effort to achieve one. The range of cases referred to and the number of sources cited were indeed impressive, even for a writer amply endowed with research assistance.

But it is obviously not the quantity but the quality of the factual evidence and the use made of it that matter. The reader with some knowledge of any part of the Third World soon finds, for example, that the use made of materials relating to that area leaves a great deal to be desired. Thus, when seeking to illustrate the rule that 'the longer a nationalist party fought for independence, the longer it was able to enjoy the power that came with independence', Huntington says that 'TANU [the Tanganyika African National Union] and its predecessor had a 32-year history when Tanganyika became independent.'[25] But TANU's predecessor (the Tanganyika African Association) could hardly be said to have 'fought for independence' before it was transformed into TANU. Or, to take a very different example from the same text, when it is a question of illustrating the functional adaptability of political institutions, the British monarchy is cited, but, when it is a question of distinguishing form from content, 'Britain preserved the form of the old monarchy, but America preserved the substance. Today America still has a king, Britain only a crown.'[26] The careful reader soon realizes that the great flood of information with which he or she is being presented cannot be taken on trust.

The central thesis — that stability is due to a high ratio of institutionalization to political participation — is supported by citing evidence from a study that compared '26 countries with a low ratio of want formation to want satisfaction and hence low "systemic frustration" and 36 countries with a high ratio and hence high "systemic frustration".'[27] Remarkably enough, one of the 26 'satisfied countries' was the Union of South Africa. The thought eventually occurs that the quality of the data used may be inferior to the quantity supplied. Formidable as it is, the volume of data too often serves precisely to dazzle, not to illuminate. Furthermore, the observant reader also notices, sooner or later, that the usual relation between the evidence presented and the hypotheses put forward is that of illustration. The effect is to make the hypotheses appear plausible, not to test them. It is difficult to find any point on which the author tries to identify cases that would constitute potentially 'disconfirming' instances for the hypothesis in question, as the textbooks on methodology prescribe.

A few attempts have been made to test the empirical validity of

[25] Huntington, *Political Order*, op. cit., p. 425.

[26] Ibid., p. 115.

[27] Ibid., p. 55.

Huntington's argument, using the kind of cross-national statistical comparisons favoured by the behavioural school to which the modernization theorists in general belong. One found a strong correlation between the ratio of social mobilization to institutionalization and political instability and violence, another found no correlation between them and a third 'unhappily reported the difficulties of selecting the correct indicators, partly due to "the thoughtful but ambiguous formulation of key theoretic concepts" by Huntington.'[28]

SCIENTISM

Of course, many other well-regarded works on politics may be faulted for poor evidential support. What makes this criticism more than usually pertinent is that Huntington's use of evidence so often takes a scientistic form. This is particularly true of the sections of *Political Order* in which the main steps of the argument are put forward. The main form of scientism is pseudo-quantification. The text abounds with propositions about 'the levels of middle class participation', 'the rate of increase of social frustration', 'the degree of corruption', and so on. This way of writing encourages acceptance of an argument by suggesting that appropriate quantitative data have been collected that verify the general statements made.

When relevant data are cited, this sometimes appears to serve to disguise a serious circularity in the argument. For instance, the statement that 'the faster the enlightenment of the population, the more frequent the overthrow of the government' is based on an alleged positive correlation between the rate of change of primary school enrolments and 'instability.'[29] The fact that there was almost certainly a similar correlation between the rate of growth of police forces and instability in the countries studied reminds us that the evidence cited only serves to support the conclusion drawn if one already accepts — on other grounds — a theory that indicates a causal connection between the rate of increase in school enrolments and revolution. Yet it is to support the plausibility of such a theory that the correlation is adduced ('the relation between social mobility and political instability seems reasonably direct').

At other times, data are cited that in no way justify the lawlike conclusion drawn. For example:

> Burma and Ethiopia had equally low per capita incomes in the 1950s: the relative stability of the latter in comparison to the former perhaps reflected the fact that fewer than 5 per cent of Ethiopians were literate but 45 per cent of the

[28] Michael C. Hudson, 'Conditions of Political Violence and Instability: A Preliminary Test of Three Hypotheses', *Sage Professional Papers in Comparative Politics* 1/5, 1920, pp. 252–53, cited in Ben-Dor, 'Institutionalization and Political Development', op. cit., p. 316, where he also refers to R. Schneider and A.L. Schneider, 'Social Mobilization, Political Institutions, and Political Violence: A Cross-National Analysis', *Comparative Political Studies* 4, 1971, pp. 69–90. The study that found no correlation is Lethinen, 'Modernization, Political Development, and Stability', op. cit.

[29] Huntington, *Political Order*, op. cit., p. 47.

> Burmese were. Similarly, Cuba had the fourth highest literacy rate in Latin America when it went communist, and the only Indian state to elect a communist government, Kerala, also has the highest literacy rate in India. *Clearly*, the appeals of communism *are usually* to literates rather than illiterates.[30]

In these examples, the quantitative form of the statement does refer to quantitative data. In other cases, however, there are no such data to refer to: the alleged relation is pure invention. For instance: 'In general, the higher the level of education of the unemployed, alienated, or otherwise dissatisfied person, the more extreme the destabilizing behavior which results.'[31]

To what kind of audience is reasoning of this sort addressed? The answer is painfully clear: it could only be addressed to the US graduate school victims of a behaviouralist *déformation professionelle* — people who have been taught to regard a statistical correlation as the epitome of proof, and not to be too concerned about the real meaning of the statistics in question, or too clear about the logic connecting any such correlation to an argument.[32]

Taken as a whole, Huntington's method is less that of enquiry than of propaganda, and, when the context is borne in mind, it strongly recalls Marx's distinction between the scientific enquiries of the classical economists and their mid-nineteenth-century successors:

> In France and England, the bourgeoisie had conquered political power. From that time on, the class struggle took on more and more explicit and threatening forms, both in practice and in theory. It sounded the knell of scientific bourgeois economics. It was thenceforth no longer a question of whether this or that theorem was true, but whether it was useful to capital or harmful, expedient or inexpedient, in accordance with police regulations or contrary to them.[33]

Huntington's work as ideology

The problem that remains to be considered is why Huntington's work, with so many serious, not to say fatal, deficiencies of logic, methodology and factual support, should have become as influential as it did.[34] The essence of the matter is this: Huntington revitalized the ideology of modernization — i.e. that partially adequate, but systematically mystified, view of the ex-colonial world that formed the basis for imperialist strategic action — in a highly significant way. First, as already mentioned, he broke the connection that 'modernization' had previously had with US pluralism. Second,

[30] Ibid., p. 49 (my emphasis).

[31] Ibid., p. 48.

[32] As is well known, the highest form of statement in the behaviouralist canon is a probability statement. Huntington's use of the term 'probability' on p. 42 of *Political Order* (op. cit.) is instructive.

[33] Postface to the second edition of *Capital*, vol. I (Harmondsworth: Penguin Books, 1976), p. 97.

[34] A survey of US professors in 1974 found that *Political Order* was the book most often cited as among the 'most important in the field'; see H.C. Kenski and M.C. Kenski, *Teaching Political Development and Modernization at American Universities: A Survey* (Tucson: University of Arizona Press, 1974), pp. 9–10.

he appropriated some of the central terms of Marxism, while simultaneously appearing to invert Marxism's tenets.

The break with pluralism was more than the mere abandonment of the democratic ideal implicit in earlier modernization writing. For Huntington, Third World societies do not consist of groups competing, in fluctuating and shifting combinations, for influence over the 'outputs' of government; nor do 'modern' systems differ from 'premodern' or 'modernizing' systems in this respect. In all political systems, the key is domination, or as Huntington says, 'power': 'Power is something which has to be mobilized, developed, and organized. It must be created.'[35] Huntington therefore wastes no time looking for the functional equivalents in the Third World of 'interest articulation' and 'interest aggregation', for example, as earlier modernization theorists did. For him, the central reality everywhere is the building of regimes capable of imposing their will on society. The basic problem of most Third World countries is that they lack such regimes. Even those who hesitated to endorse this latter view (especially since Huntington thought this was the USA's problem too) could see that his conception of politics was much better adapted than the pluralist model to the conflictual realities of the Third World, from General Geisel's Brazil to Indira Gandhi's India — or, indeed, Castro's Cuba.

This brings us to the relation between Huntington and Marxism. In considering this, it must be borne in mind from the outset that Huntington's theory was not coincidentally, but directly and intimately, connected with the war in Vietnam. From 1966 to 1969 he was chairman of the Council on Vietnamese Studies of the US Agency for International Development's South-East Asia Advisory Group, and in 1967 he visited Saigon on behalf of the State Department 'to investigate ways in which political power could be developed in Vietnam'.[36] The following year — the year in which *Political Order* was published — he wrote an article in which he explicitly interpreted the Vietcong's success as due not to its ideals but to its ability to impose authority in rural areas where authority was lacking:

> There is little evidence to suggest that the appeal of the Viet Cong derives from material poverty or that it can be countered by material benefits...The appeal of revolutionaries depends not on economic deprivation but on political deprivation, that is, on the absence of an effective structure of authority.[37]

According to Huntington, the authority of the Vietcong in the rural areas was unlikely to be overthrown, but, on the other hand, the US entry into the war in 1965 had led in the following three years to about 3 million people fleeing the countryside for the cities, where they came under the authority of Saigon. Huntington considered that, as a result, the possibility

[35] Huntington, *Political Order*, op. cit., p. 144.

[36] John Gretton, 'The Double-Barrelled Character of Professor Huntington', *Times Educational Supplement* 29, June 1973, p. 10.

[37] S. Huntington, 'The Bases of Accommodation', *Foreign Affairs* 46, 1968, p. 644.

of successful revolution in South Vietnam was being undercut because 'the good Maoist expectation that by winning the support of the rural population, it [the Vietcong] could eventually isolate and overwhelm the cities' was being invalidated by 'the American-sponsored urban revolution'.[38] The answer to wars of national liberation was, therefore, 'forced-draft urbanization and modernization which rapidly brings the country in question out of the phase in which a rural revolutionary movement can hope to generate sufficient strength to come to power'.[39]

If this article revealed starkly the political thrust of the 1968 book, it is equally important to stress that defeating the revolution in Vietnam did not, for Huntington, necessarily entail defeating the Vietcong. On the contrary, he accepted that the Vietcong's power in the rural areas could not in practice be destroyed, and he proposed that a negotiated peace settlement be based on this fact, and that the Vietcong-National Liberation Front (VC-NLF) should be encouraged to 'enter the political process' through participation in electoral politics, at first locally and eventually on a national level: 'If as a result of this process the VC-NLF secured control of the Central Government, the United States would obviously regret the outcome but could also accept it and feel little compulsion to re-intervene.'[40]

In other words, Huntington carried his antirevolutionary outlook to its logical conclusion by separating it from anticommunism. For him, it was the ideal of a fundamental reconstruction of society that was pernicious, because utopian and destructive of order.[41] He did not see the Vietcong as a revolutionary force in this sense, but as a force potentially capable of imposing order. The lesson that Third World countries such as Vietnam needed to learn — how to build and impose authority — could, he said, be learned in Moscow and Peking and not in Washington.[42] Indeed, far from being anticommunist, Huntington expressed almost unqualified admiration for the political regimes of Stalin and Mao, criticizing Stalin only for weakening the party from 1936 onward and Mao for launching the Cultural Revolution.

This practical insight — that it was not communism, not even as embodied in the Vietcong, that was the real enemy, but revolution, to which the bureaucratized Communist parties were equally opposed — was what Huntington's book developed ideologically, by a double process. First, Marxism — the theory and practice of the emancipation of labour — was treated

[38] Ibid., p. 650.

[39] Ibid., p. 652. Perhaps these events are already sufficiently remote for it to be necessary to remind the reader that the 'American-sponsored urban revolution' and 'forced-draft urbanization and modernization' refer to the saturation-bombing and scorched-earth techniques by which the USA and the Saigon regime attempted to deny the countryside to the Vietcong, at an estimated cost of 2 million killed or wounded and 8 million refugees — out of a total South Vietnamese population of some 16 million.

[40] Ibid., pp. 655–56.

[41] Huntington, *Political Order*, op. cit., p. 371.

[42] Ibid., pp. 137–38.

dismissively, as a partial theory ('Marx focussed on only one minor aspect of a much more general phenomenon') and as politically naïve ('Lenin was not a disciple of Marx, rather Marx was a precursor of Lenin...Marx was a political primitive').[43] In place of Marx, Huntington elevated Lenin and 'Leninism' — which, however, turned out to be Stalinism: 'Marxism, as a theory of social evolution, was proved wrong by events; Leninism, as a theory of political action, was proved right...Lenin laid down the prerequisites for political order...In Leninism the party is not just institutionalized; it is deified.'[44] Very few of Huntington's US readers would know the difference; one cannot be sure that Huntington knew it himself. The important point was, in any case, that 'Marxism' was wrong.

Second, in 'accepting' Leninism (read: Stalinism) simply as an up-to-date and efficient doctrine of authority, Huntington was able to accept (or rather to appear to accept) what most previous modernization theorists had been at pains to ignore — social classes and revolutionary movements — as central phenomena of Third World politics. In effect, Huntington seemed to be taking over the well-known Marxist concepts for dealing with Third World realities while dispensing with — or even inverting — the revolutionary theory in which they were normally contained. Moreover, Huntington demonstrated how this could be done without making any significant change in the essential structure of the ideology of modernization. What this implies is that what he appropriated from Marxism was not its concepts but only a selection of the connotations of the terms denoting these concepts — connotations which he then attached to different concepts, compatible with the rest of the ideology.

The essence of this process can be illustrated from virtually any part of *Political Order*. Take social classes. For Marx, social classes were the prime movers of history, formed through struggles within and over specific relations of production inherited from the past. The analysis of the formation, organization, alliances, defeats and victories of revolutionary classes and their antagonists, and of the gradually developing consciousness of their interests and historical tasks, constituted for Marx a tightly bounded and intricate theoretical and empirical task, with all-important implications. In contrast, Huntington's concept of class entails no such consequences. The classes or groups he writes about (he more often uses the latter term) are not considered as the products (let alone the makers) of determinate historical relations of production. They appear as empirically 'given' universal categories (though a somewhat mixed bag — lumpenproletariat, industrial workers, middle-class intelligentsia, landowners, peasants), which are assumed to exist more or less everywhere. They are no more than occupational or socio-economic aggregates, mere reservoirs of potentially 'participant' individuals, who may or may not be 'mobilized' by 'élites'. Huntington's Third World may thus seem to contain social classes, but they

[43] Ibid., pp. 37 and 336.
[44] Ibid., pp. 337–39.

are conceived in such a wholly ahistorical, unanchored, empiricist way that virtually anything may be said about them without excessive fear of contradiction.[45]

This enables Huntington to appear to accept Marxist concepts while simultaneously 'disproving' Marxism. An obvious example is his discussion of the 'lumpenproletariat', a term taken directly from Marx, but referring in Huntington merely to the newly immigrant urban 'poor' — a concept as distant from Marx's as it is vague. Huntington's main point about this 'lumpenproletariat' is that it is 'on the surface, the most promising source of urban revolt', but that, in fact, in the 1960s it did not prove to be so.[46] Probably most of Huntington's readers would have been surprised to know that: (i) Marx's concept of the lumpenproletariat had very little to do with the urban 'poor'; and (ii) Marx considered the lumpenproletariat to be the natural ally of reaction, and Engels made non-co-operation with the lumpenproletariat a point of political principle for the working class.[47] In other words, Huntington, by appropriating Marx's term but attaching it to an altogether different concept, was able to give the appearance of sophistication and realism, and to seem to be 'disproving' Marxism, while in reality attacking a not very plausible straw man. Huntington's treatment of 'industrial workers' (the proletariat?), the 'middle classes' (petty bourgeoisie?), and the 'peasantry' proceeds in the same fashion. The reduction of the Marxian concept of 'classes' to the concept of mere 'economic groups' permits discussion to be divorced from any consideration of the actual historical conditions in which they have been and are developing, the actual course of their struggles, the role of imperialism in those struggles, etc.

In effect, Huntington introduces 'classes' into the ideology of modernization simply as forms in which the masses threaten the maintenance of order; and their capacity to do so, in his formulation, is already given by the 'institutionalization' of the 'polity', i.e. in practice, by the capacity of the 'regime' (which never has a class character) to co-opt, deflect or repress this threat. As Henry Bernstein aptly remarks:

> The determining model of modernity, from which everything else follows, is itself non-problematic as it is already 'given' by the historical development of the West. This mode of conceptualization can only produce answers that are already determined by the way in which questions are posed.[48]

Similarly with 'revolution'. Marx's concept is historical, and carries a very specific socio-economic content; for him, a revolution is always a specific process of transition through which a particular country or region

[45] Huntington's use of 'class' is evidently closer to Weber's than to Marx's, but is not rigorously based on Weber's usage either.

[46] Huntington, *Political Order*, op. cit., pp. 278–83.

[47] F. Engels, 'Prefatory Note to The Peasant War in Germany', K. Marx and F. Engels, *Selected Works*, vol. I, p. 646.

[48] Henry Bernstein, 'Sociology of Development vs. Sociology of Underdevelopment?', in D. Lehmann (ed.), *Development Theory: Four Critical Essays* (London: Frank Cass, 1979).

passes, from an epoch dominated by one mode of production (such as feudalism or capitalism) to a new epoch dominated by a new mode of production. For Huntington, however, 'revolution' refers (as we have seen) simply to any 'rapid, fundamental, and violent change'. For historical materialism, the question of whether a revolution can or will occur in a given society at a given moment, and what its significance or historical 'content' may be, is always a problem of analysing the development of the contradictions to which the existing mode of production gives rise and their expression in class struggles, whereas, for Huntington, revolutions are merely pathological modes of restoring order. Thus, instead of Marx's concept, embedded in a theory for assessing the prospects for particular class alliances to make particular transformations of particular societies, Huntington's concept makes revolutions mere aberrations in the global march toward the mass consumption society to which the whole of modern history is ultimately reduced by modernization theory.

In conclusion

Huntington thus revitalized the ideology of modernization in several ways at once. He offered a crude but substantive phenomenology of classes/groups, power struggles, revolutions, corruption, militarism and the like, important parts of which seemed to be taken over from a simultaneously disparaged Marxism, and all of which could be — in the form given by Huntington — integrated into the ahistorical and protean schema of tradition–modernity. As scientific theory it was false. As propaganda it was crude, although not necessarily ineffective, especially within the closely patrolled intellectual confines of most US political science departments. But as a model of ideological reconstruction it was path-breaking, a dramatic example of the 'ideological flexibility of which bourgeois thought is capable', and one which was to have a profound influence on bourgeois thinking about development in the 1970s.[49]

[49] Ibid., 'Conclusion': Bernstein's penetrating essay should be read by everyone interested in the process of construction and reconstruction of bourgeois social science.

Rational Choice or Hobson's Choice? The 'New Political Economy' as Development Theory

The adoption of neo-liberalism as the ruling philosophy of the World Bank and the imposition of 'structural adjustment' programmes on Third World countries in the name of this philosophy were not something to which political scientists made much of a contribution; credit for the Bank's conversion, in particular, may be given almost entirely to economists. On the other hand, political scientists have not been conspicuous in criticizing it either; on the contrary, in the USA, where most political scientists are to be found, neo-liberal thought has had some support from the growing vogue for 'rational choice' theory in political science and related disciplines from 'management' to history.

Directly, the work of Robert Bates, a political scientist who applied a rational choice-based kind of 'political economy' to the analysis of the crisis in African agriculture, was taken up explicitly by the Bank;[1] so was the thinking of the economic historian Douglass North, who argued in 1981 that what he called the 'new institutionalism' promised to yield a general theory of development from the earliest times to the present.[2] And by 1994 the Bank devoted its annual *World Development Report* to the theme of infrastructure provision in terms almost wholly derived from the work of a network of 'new institutionalists' working on middle-level policy issues.[3]

[1] The seminal work was Robert H. Bates, *Markets and States in Tropical Africa* (Berkeley: University of California Press, 1981).

[2] The work for which North shared a Nobel Prize in 1993 was in (primarily American) economic history. His first and most influential attempt to sketch an institutional theory of development was *Structure and Change in Economic History* (New York: W.W. Norton, 1981).

[3] The World Bank's *World Development Report* for 1986 (Washington: World Bank), which focused on agricultural policy, adopted Bates's analysis without qualification and summarized it virtually word for word on p. 92. The 1987 *Report*, focusing on foreign trade and industrialization, relied explicitly on North and cited his comparison of the contrasting development paths of the UK and Spain (p. 61) as support for its emphasis on the need for governments to concentrate on establishing clear (private) property rights rather than involve themselves in productive activities. The 1991 *Report*, on 'The Challenge of Development', relies heavily on a blend of Bates and North to buttress its core neo-liberal

It is true that the new theory of development envisaged by North has not emerged, nor has Bates's work in Africa yet had many imitators, at least among students of politics.[4] On the other hand, a kind of partial hegemony of rational-choice ideas has undoubtedly been occurring in US political science, and not least in the field of development, as can be seen in the constant citation of Bates and North (and some forerunners who are enjoying a belated celebrity), and in the increasing use of the terms, if not the concepts, of the 'new political economy/institutionalism' in the discourse of non-economists: expressions like 'transaction costs', 'rent-seeking' and 'free rider' appear peppered throughout it, if not always in a particularly illuminating fashion.[5] Bates has always

arguments: for particular instances see pp. 10–11 and 145. The debt of the 1994 *Report* on infrastructure to the choice-theoretic literature is explicit, and, in so far as it is restricted to the technical implications of alternative institutional structures for providing a range of different services, less objectionable, although the *Report's* treatment of the issues at stake as if they were purely 'technical', by excluding crucial political and social considerations, is patently ideological.

[4] See p. 1036 of Howard Stein and Ernest J. Wilson III, 'The Political Economy of Robert Bates: A Critical Reading of Rational Choice in Africa', *World Development* 21/6, 1993, pp. 1035–53. Various political scientists have, of course, made a particular concept drawn from the theory of choice, especially the concept of 'rent-seeking', the organizing principle of their research: see for instance Catherine Boone, *Merchant Capital and the Roots of State Power in Senegal, 1930–1985* (Cambridge and New York: Cambridge University Press, 1992); see also several of the contributions to Jennifer A. Widner (ed.), *Economic Change and Political Liberalization in Sub-Saharan Africa* (Baltimore: Johns Hopkins University Press, 1994). Mark Gallagher, an economist at the World Bank, has tried to analyse the costs of rent-seeking in Africa in *Rent-Seeking and Economic Growth in Africa* (Boulder: Westview Press, 1991), but no one seems to have advanced the general theory proposed by Bates or applied it to any other aspects of African development. An anthropological study which emulates Bates's later fieldwork on Kenya, using the perspective of the new institutionalism to probe the reactions of Kenyan cattle-herders to relative price changes, is Jean Ensminger, *Making a Market: The Institutional Transformation of an African Society* (Cambridge: Cambridge University Press, 1992).

[5] An interesting example of this in the development field is Goran Hyden's introductory chapter, 'Governance and the Study of Politics', in Goran Hyden and Michael Bratton (eds), *Governance and Politics in Africa* (Boulder: Lynne Rienner, 1992), especially pp. 5–12, which displays some unease towards the choice-theoretic literature while at the same time citing it extensively in the construction of a concept of 'governance' (defined as 'the conscious management of regime structures with a view to enhancing the legitimacy of the public realm'). What is most notable about these references is that at no point is any substantive theoretical argument contained in this literature even mentioned, let alone discussed or endorsed; it is as if the mere citation of the literature adds legitimacy to Hyden's project. Sometimes the concept of 'choice' seems to function, along with 'diversity' (which often seems to carry an implicit endorsement from the literature of post-structuralism and post-colonialism), simply as a sort of universal solvent, melting down all clear distinctions and oppositions drawn from pre-existing schools of thought. The following example is, unfortunately, not exceptional: 'Politics', Naomi Chazan and her colleagues, introducing their textbook, explain, 'are...perceived as a set of transactions, the manifestation of the exercise of choice by multiple actors within existing parameters. This perspective, unlike the managerial view of the modernization school, the exploitative emphasis of dependency approaches, and the instrumental notions of statist writers, highlights the fluidity of politics and attempts to trace the vacillating [*sic*] political course' (Naomi Chazan *et al.*, *Politics and Society in Contemporary Africa* (Boulder: Lynne Rienner, second edition 1992), p. 23.

distinguished his 'political economy' from 'market-based reasoning';[6] but his reputation as an unusually accomplished interdisciplinary thinker who has made rational choice his chosen approach to the problems of African development and North's reputation as a Nobel prize-winning rational-choice theorist of development have helped to foster a somewhat uncritical respect for rational-choice ideas among US political scientists concerned with development, which has lent the neo-liberals in the 'development community' a degree of intellectual respectability among non-economists that they badly needed.

Yet the considerable influence of rational-choice theory in US political science generally has not been established purely on its merits.[7] Its main achievements have been in throwing light on relatively micro-level institutional problems; attempts to present it as a new general theory of politics have been singularly unsuccessful. So its currently influential status is curious. How, then, should we account for it? To explore this question in relation to the study of development, we need to begin with a brief look at the genesis of the 'new institutionalism' and the 'new political economy' (this can be skipped by readers who are familiar with it).

Sources of the 'new political economy'

Although there is not yet much agreement about labels among the practitioners of rational choice-based work on institutions, we can distinguish three broad categories, at different degrees of distance from neo-classical economics. First there is the 'new institutional economics' (NIE), a branch of economics. Bardhan distinguishes two streams within it.[8] One, the closest

[6] See, for example, R.H. Bates, 'Macropolitical Economy in the Field of Development', in James E. Alt and Kenneth A. Shepsle (eds), *Perspectives on Positive Political Economy* (Cambridge: Cambridge University Press, 1990), especially pp. 38–41 and 51–54. This paper includes an illuminating account of Bates's shift from early modernization theory to rational choice without, apparently, any significant encounter with Marxist political economy along the way, although Bates, like North, thinks there are affinities between the agenda and approach of Marxism and his own ultimate standpoint.

[7] For incontrovertible evidence on this point see the very balanced discussion by Donald P. Green and Ian Shapiro in *Pathologies of Rational Choice Theory: A Critique of Applications in Political Science* (New Haven: Yale University Press, 1994), where they show both the dramatic rise of rational-choice theory in US political science (accounting for nearly 40% of all articles in the *American Political Science Review* by 1992) and its acute shortcomings, concluding: 'If our assessment of rational choice scholarship is on the mark, it would not be the first time that a theory widely accepted at one time is later revealed as mistaken or oversold. Aristotle's physics, the view that the earth is flat or that it is at the center of the universe, and Lamarckian genetics are all cases in point' (p. 195). I regret that this excellent book was not available when I was doing most of the work for this chapter; on the other hand, it is interesting that Green and Shapiro make no reference to the work of Bates or any other political scientist who has applied rational-choice theory to the study of development.

[8] Pranab Bardhan, 'Alternative Approaches to the Theory of Institutions in Economic Development', in P. Bardhan (ed.), *The Economic Theory of Agrarian Institutions* (Oxford: Clarendon Press, 1989).

to neo-classical economics in its concern for formal rigour, is based on the theory of imperfect information; its most-cited founding text is George Akerlof's 'The Market for Lemons'.[9] As Bardhan and others have demonstrated, this body of theory has many useful applications in development, especially for relatively long-term, stable agricultural policy areas, from tenancy to labour contracts and credit systems. The other strand distinguished by Bardhan is focused on property rights and transaction costs, starting from an article published by Coase in 1937 on the reasons for the existence of firms:[10] Bardhan calls it the CDAWN school, after Coase, Demsetz, Alchian, Williamson and North.[11]

Douglass North, however, does not altogether belong in this company. Coase, Demsetz, Alchian and Williamson are economists who have all been primarily concerned with explaining the existence, internal organization and strategic behaviour of firms. North, in contrast, is a historian concerned to make explanations in economic history compatible with explanations in contemporary economics, i.e. to bring them within the conceptual framework of neo-classical theory. In pursuit of this aim, he has pioneered a second current of thought known to its practitioners as 'the new institutionalism'. The writers in this current belong to various disciplines and are concerned with a variety of middle-level problems; many of them are policy-orientated.[12] They aim to apply the findings of all the various currents of rational-choice theory to institutions of all kinds, from management or fiscal policy to infrastructure provision and irrigation schemes.

[9] G. Akerlof, 'The Market for Lemons: Qualitative Uncertainty and the Market Mechanism', *Quarterly Journal of Economics* 84/3, 1970, pp. 488–500.

[10] Ronald Coase, 'The Nature of the Firm', *Economica* 4, 1937, pp. 386–405.

[11] Bardhan acknowledges that information costs, the core concept of the Akerlof-Stiglitz school to which he and his colleagues mainly subscribe, are a sub-category of transaction costs. Although some writers see transaction costs and property rights as constituting separate lines of analysis, most, like Yoram Barzel, seem to agree with Bardhan that they belong together: 'The concept of property rights is closely related to that of transaction costs. I define transaction costs as the costs associated with the transfer, capture and protection of rights' (Yoram Barzel, *Economic Analysis of Property Rights* (Cambridge: Cambridge University Press, 1989), p. 2). For a helpful survey of the 'CDAWN' school, and a selection of their seminal articles, see Louis Putterman (ed.), *The Economic Nature of the Firm* (Cambridge: Cambridge University Press, 1990).

[12] Elinor Ostrom's study (*Governing the Commons* (Cambridge: Cambridge University Press, 1990)) of the effects of alternative ways of organizing access to and maintenance of common property (grazing, water, fish stocks) is an influential text here; see also Elinor Ostrom, Larry Schroer and Susan Wynne, *Institutional Incentives and Sustainable Development: Infrastructural Policies in Perspective* (Boulder: Westview, 1993). The series in which both these books are published, The Political Economy of Institutions and Decisions, edited by Douglass North and James Alt, is a useful index of the scope and focus of the new institutionalism. Besides several policy-orientated studies of organizations like Ostrom's, the books probe the power of institutional analysis to illuminate a variety of situations, from 'a century of Swedish politics' (the sub-title of Leif Lewin's *Ideology and Strategy* (Cambridge: Cambridge University Press, 1988)) to (more typically) 'The Design of the Appropriations Process in the House of Representatives 1865–1921' (the sub-title of Charles Stewart III's *Budget Reform Politics* (Cambridge: Cambridge University Press, 1989)).

The third branch of rational-choice work on institutions, is Bates's 'new political economy' (NPE).[13] It differs from the new institutionalism mainly in its sensitivity to politics, and in the qualifications that it is, as a result, ready to make to the assumptions of choice-theoretical analysis. Douglass North himself really belongs in this stream, although he does not describe himself as a political economist; because he wants to explain long-term as well as short-term change, and at the national and even global level, he too has to confront the problem that politics poses. And, unlike the more policy-orientated among the new institutionalists, he is interested not only in what productive or distributional effects particular institutional arrangements may have, but also in how existing institutions come into being, and how they change over the long term; and he is perhaps right in thinking that a general theory that comprehended all of this would be a general theory of development.

The first question is, then, whether the NPE, represented by Bates explicitly and by North in practice, demonstrates the potential they, to differing degrees, attribute to it. As the NPE rests so heavily on the insights of the NIE and the new institutionalism, we need first to take a brief look at the sort of insights these are.

The new institutional economists start out from the theoretical framework of neo-classical economics, in which markets are frictionless: any change in supply or demand at any point in the system leads automatically to corresponding changes throughout the system until equilibrium is restored and no further change anywhere in the system will increase the total sum of satisfaction of all the individuals in it. But the new institutional economists observe that in reality information is always imperfect and/or that market actors do not respond to price changes automatically but do so through a vast web of transactions that have costs (including the cost of getting better information). Transactions take place in institutionalized ways; institutions are systems of rules which to a greater or lesser extent reduce these costs, i.e. they may be more or less efficient in terms of Pareto optimality. The new institutional economists analyse the logic of these institutions within the assumptions of neo-classical economics, i.e. they study the results of the way different institutionalized rules affect the

[13] Bates calls what he does simply 'political economy', although he sees it as a discipline still in the process of formation (as in the title of his edited volume, *Toward a Political Economy of Development: A Rational Choice Perspective* (Berkeley: University of California Press, 1988). Bates's usage is quite close to that of Popkin, who called his rational-choice-based explanation of peasant revolutionary activity in Vietnam 'political economy' to distinguish it from the (also rational-choice-based) 'moral economy' explanations of James Scott and others (see Samuel Popkin, *The Rational Peasant: The Political Economy of Rural Society in Vietnam* (Berkeley: University of California Press, 1979)). Other writers are at pains to use the term 'new political economy' to differentiate it from the old 'radical', 'Marxian' or 'neo-Marxist' political economy that was influential, in development theory at any rate, in the 1970s, rather as the 'new institutionalists' differentiate themselves from 'old institutionalists', who were essentially historians, not rational-choice theorists.

behaviour of rationally choosing individuals seeking to maximize their given material welfare preferences.

North explains the general starting-point with a simple example: how does he know, when buying oranges, if he is getting good ones?[14] It would cost too much time and money to make sure that each and every purchase was up to standard; instead, he trusts in his knowledge of the reliability of the particular greengrocer he gets them from. Even this knowledge is not acquired costlessly, but the costs of knowing whether the product is satisfactory become far greater if it is only an occasional purchase and from a supplier with whom he has no personal acquaintance; and the more complex and impersonal markets become, the more typical this situation is. Yet the benefits of specialization and economies of scale mean that markets must get more complex and more impersonal; consequently all sorts of non-market institutions come into existence to reduce these costs. The state sets grading standards and employs inspectors to check them; it establishes courts in which to prosecute dishonest sellers, and through which contracts that specify quality can be enforced. (In the same way, Coase argued, firms themselves can be understood as a means of reducing transaction costs by bringing a whole series of transactions inside one set of buildings and subordinating them to oversight by a centralized hierarchy of supervisors.) In general, the state establishes property rights, spreads risks (e.g. through state-provided export credit guarantees) and allocates the costs of 'externalities' (e.g. the costs of factory-produced pollution).

Most institutions, from kinship systems to religions, are 'organic', i.e. they have evolved gradually under many influences over very long periods; but many of them affect production and distribution and they too can be analysed in rational-choice terms. On the whole, however, it is 'pragmatic', or purpose-built, economic institutions that the new institutional economists and the new institutionalists study.[15]

The analysis proceeds with a very limited stock of core concepts to which real-life phenomena are reduced. Institutions are any set of rules governing behaviour. Individuals may be principals or agents (or both; for example, as voters they are principals and their elected representatives are their agents, while as workers they are agents and their employers are principals). They

[14] North, *Structure and Change in Economic History*, op. cit., pp. 34–36.

[15] A significant exception to this is Gary Becker's *A Treatise on the Family* (Cambridge: Harvard University Press, 1981), where a wide range of family patterns (family size, divorce, and so on) are explained in terms of the assumptions of neo-classical economics. See also his 'The Economic Approach to Human Behaviour', in Jon Elster (ed.), *Rational Choice* (Oxford: Blackwell, 1986), pp. 108–22, in which he concludes (p. 19) that 'human behaviour is not compartmentalized, sometimes based on maximizing, sometimes not, sometimes motivated by stable preferences, sometimes by volatile ones, sometimes resulting in an optimal accumulation of information, sometimes not. Rather, all human behaviour can be viewed as involving participants who maximize their utility from a stable set of preferences and accumulate an optimal amount of information and other inputs in a variety of markets. If this argument is correct, the economic approach provides a unified framework for understanding behaviour that has long been sought by and eluded Bentham, Comte, Marx and others.'

may act alone or in groups. Groups may be large and dispersed, or small and concentrated. Everything people want is a 'good'. Everything they don't want to do or must give up to get something they want is a 'cost'.

Such simplifications allow the new institutionalists to identify, and in some cases to model mathematically, the logic of rule-governed behaviour that may be relevant to widely differing settings, and so to formulate principles that may have quite general application. Examples of the kinds of issues they study have become familiar, at least by their labels. 'The free-rider problem' refers to the question of how collective action can be achieved to secure a 'public' or 'non-excludable' good (e.g. something like a clean water supply, or clean government), which no one can really be prevented from enjoying, regardless of whether they have helped to bring it about. 'The prisoner's dilemma' refers to the problem of the sub-optimal choices made by people in situations where they cannot know each other's minds, and where there are potential advantages to be gained by any one of them failing to co-operate while the others continue to do so. 'The tragedy of the commons' refers to the destruction of resources that can occur if people have common property rights in something like fish stocks or grazing land, and to alternative systems of property rights that avoid this, and their various productive and distributive effects. 'Moral hazard' refers to problems that arise by analogy with insured persons or groups who, once they are insured, may change their behaviour in ways that raise the risk to the insurer. 'Opportunism' refers to the various ways of dealing with the fact that people have an incentive to cheat, or at least not to live fully up to their moral obligations, in situations where it is costly to supervise and/or penalize them.

Problems of the 'new political economy'

But the extension of this body of thought to include questions about the reasons for the formation of institutions, as well as about their effects, means introducing politics, and this poses serious theoretical problems, as both Bates and North have recognized.

First, some institutions are hard to explain except as the product of decisions taken in the name of some general concept of the social welfare; and this means abandoning the neo-classicists' commitment to grounding all analysis on the behaviour of individuals rationally pursuing their purely individual interests.[16] A concept of the polity, or society as a whole, is also

[16]This is true, Bates thinks, even of the 'new institutionalists', in so far as they seek to account for the origins of institutions as a response to the shortcomings of market transactions as they would be in the absence of these institutions: for them, 'it is the needs of society that call forth non-market organisation' (Robert H. Bates, 'Social Dilemmas and Rational Individuals: An Assessment of the New Institutionalism', in John Harriss, Janet Hunter and Colin Lewis (eds), *The New Institutional Economics and Third World Development* (London: Routledge, forthcoming 1995), draft p. 30. I am grateful to Professor Bates for allowing me to read an early draft of this chapter). But many non-market organizations can be supposed to have been formed to respond to the needs of the individuals or groups who promoted their formation, even though they are typically represented as the needs of 'society'.

indispensable for assessing the long-term political and social implications of institutional change; and issues such as inequalities of political power, the non-economic goals of politicians and the role of ideology, which seem inescapably part of this story, cannot be boiled down to the parsimonious categories of the conceptual framework of neo-classical economics without a self-defeating loss of meaning.

Secondly, the assumption of individual rationality is also abandoned to the extent that institutions that were designed to serve the interests of 'society as a whole' are by definition themselves 'public' goods, so that getting them established must always involve overcoming the free-rider problem. But in many situations this is not possible without relaxing the assumption that rational individuals only act so as to maximize their individual material welfare (i.e. no special incentives can be offered by identifying some additional 'excludable' benefits to be given only to those who take part in the collective effort), so some other assumptions are required for explaining how institutions do in fact get changed.

The first problem is clearly fundamental. Neo-classical economists do not think that there is any conflict between the goal of achieving a Pareto-optimal satisfaction of people's present individual preferences and the goal of maximizing the growth of total production in the long term. But if, like Bates and most, if not all, observers of politics, you think these goals can conflict, then either you must regard individuals as having conflicting preferences (e.g. between short-run and long-run goals), which makes it impossible to model their behaviour on the assumptions of rational-choice theory; or you must posit a conflict between their individual preferences and some other set of preferences, however derived, that belong to them collectively, which also cannot be accommodated within the assumptions of rational-choice theory. The idea that rational-choice theory can govern the whole explanatory enterprise must be abandoned, and it then becomes an open question how far it can usefully govern it at all.

And the second problem is no less fundamental. Institutions that serve the general interests of 'society' cannot be explained in terms of their serving this interest if the logic of rational choice would always tend to prevent their being established. (One rational-choice theorist, Jack Knight, is led by this line of reasoning to a conclusion with a distinctly Marxist flavour: '[collective benefits] are merely by-products of distributional conflict').[17]

These two difficulties mean that the new political economy cannot be a 'pure' mode of enquiry.[18] It cannot restrict itself to assumptions and axioms that are derived from, and that would permit it to be integrated with,

[17] Jack Knight, *Institutions and Social Conflict* (Cambridge: Cambridge University Press, 1992), p. 20.

[18] By 1990, if not earlier, Bates seemed to claim for rational-choice-based analysis only the status of a complementary tool alongside structural and cultural approaches, including Marxism (see 'Macropolitical Economy', op. cit., pp. 53–54). His own work, however, has not so far really used it in this way.

neo-classical economics or rational-choice theory in general. To deal with its subject matter it must abandon the 'rigour' and 'parsimony' that constitute such an important part of its appeal. It becomes another heuristic tool, philosophically and empirically on a par with others — on which basis, moreover, it should not necessarily be unwelcome.[19]

In practice, however, the 'new political economists' rarely seem to operate as if this were the case; they express a distinctive methodological preference. At least two rather basic theoretical contradictions are acknowledged, and a very eroded conception of politics is preferred to the much richer and more subtle *problématiques* that are available, so that the general method of rational choice analysis can be transposed — even if only partially — from economics to the political realm. These liabilities are accepted because of the supposed explanatory power of the insights of the original body of theory. The question is, then, how far the sacrifices are worth while. The next two sections consider this in terms of Bates's and North's work respectively.

The NPE as macro-level political economy: Robert Bates

Robert Bates won attention for his ideas through three books on Africa published in the 1980s.[20] Although his 1987 volume of essays and his country case-study of Kenya applied choice-theoretic ideas to a variety of themes, the central thesis, with which he started and finished, was that the reason why African governments generally kept down the produce prices paid to farmers, and why they tried to increase food supplies not by raising these prices but by costly and inefficient state-run agricultural projects and subsidies for farm inputs, was that this was politically rational from the point of view of African governments, if irrational from the point of view of the people's collective welfare. The collective welfare would have been best served by a policy of raising prices for farm produce. But most farmers are small-scale and hence numerous and hard to organize. Large-scale farmers are few and hence more readily organized, and they benefit from subsidies on their inputs; urban workers are concentrated and so also more easily organized, and they benefit from cheap food. Also, state agricultural projects create patronage and profit opportunities for government supporters. And all these policies, including farm prices administered by state marketing boards, generate rents which politicians and officials can appropriate for

[19] I should perhaps make it clear here that I myself consider rationality (and, in the weak sense defined by Jon Elster in his *Making Sense of Marx* (Cambridge: Cambridge University Press, 1985), methodological individualism as well) a necessary starting-point for political analysis. I am only sceptical about how far we actually get in understanding most important political problems with the 'insights' yielded by the particular version of rationality on which rational-choice-based theory is built.

[20] R.H. Bates, *Markets and States in Tropical Africa*, op. cit.; *Essays on the Political Economy of Rural Africa* (Berkeley: University of California Press, 1987); and *Beyond the Miracle of the Market* (Cambridge: Cambridge University Press, 1989).

themselves or use as patronage. Therefore these policies are rational for those in power, and this explains them.

Bates's thesis has been criticized on various grounds, which can perhaps be summarized as follows.

1. In representing low producer prices as self-evidently contrary to the collective interests of society it occludes some of the crucial determinants of the social misuse of agricultural surpluses.[21]

2. It posits an equilibrium when it is apparent that economic policy has been anything but the result of an equlibrium.[22]

3. While purporting to make no assumptions about the values individuals maximize, in practice it always assumes that what individuals try to maximize is their income, whereas this is often not a plausible assumption (they may seem, and even claim, to give priority to enhancing their personal security, social status, political status, state of grace, etc).[23]

4. It relies on 'stylized facts', rather than a systematic analysis of comparative evidence to support its conclusions.[24]

Of these objections it is the fourth that will be chiefly considered here. The first objection is valid: Bates does build into his model a preference for the market rather than the state as an allocator of the social surplus. But if this misleads him it should be clear that this is so; the merit of the highly simplified models of the NPE should be precisely to be capable of showing, by virtue of their simplicity, just what can and what cannot be explained in terms of the model.[25] Similarly with regard to point 2, an equilibrium model is just a model; using it does not commit us to the view that the real world is in equilibrium, or even that it tends towards equilibrium.[26] All it can do,

[21] Manfred Bienefeld lists the following 'excluded and neglected problems': the possibility that long-run efficiency may not be served by short-run market allocation of resources; changing world market conditions; the role of foreign investors and policy advisers ('Analysing the Politics of African State Policy: Some Thoughts on Robert Bates's Work', *IDS Bulletin* 17/1, 1986, pp. 5–11. Stein and Wilson ('Political Economy of Robert Bates', op. cit.) criticize Bates for disregarding the role of non-economic considerations on the part of policy-makers and the role of exploitation and domination as part of the incentive structure. Sara Berry considers that Bates's analysis overlooks the degree to which policies and laws fail to determine behaviour, and the degree to which individuals systematically choose not to choose, i.e. to keep open as many options as possible ('Understanding Agricultural Policy in Africa: The Contributions of Robert Bates', *World Development* 22/6, 1993, pp. 1055–62).

[22] This point is forcefully made by Pauline Peters on pp. 1071–72 in 'Is "Rational Choice" the Best Choice for Robert Bates? An Anthropolgist's Reading of Bates's Work', *World Development* 21/6, 1993, pp. 1063–76.

[23] Stein and Wilson, 'Political Economy of Robert Bates', op. cit., pp. 1042–44.

[24] Ibid., p. 1048.

[25] A similar general defence of Russell Hardin's and Samuel Popkin's use of rational choice theory is put forward by Brian Barry in 'Methodologies vs. Ideology: The "Economic" Approach Revisited', in Elinor Ostrom (ed.), *Strategies of Political Enquiry* (Beverley Hills: Sage, 1982), pp. 123–47.

[26] '"Models" — I frequently feel forced to cite [Henri] Theil — "are to be used, not believed"' (Adam Przeworski, *Democracy and the Market: Political and Economic Reforms in Eastern Europe and Latin America* (Cambridge: Cambridge University Press, 1991), p. 30).

as Bates insisted in reply to his critics, is suggest what would happen if the relationships in the model held in reality, allowing us to identify and understand more clearly, to the extent that the real world does not conform to the model, why it doesn't.[27] And, however unrealistic it may be to posit that individuals always give priority to their economic interests (point 3), if this assumption predicted (or retrodicted) people's behaviour well, it would have earned its keep.

Objection 4, on the other hand, is for the same reason a vital one. If the artificiality of the model is justified by its potential explanatory power, the extent to which it really does explain becomes crucial. And here it must be said that Bates does not adopt either of at least two possible methodological strategies that his approach would seem to call for. He does not undertake a systematic comparison of a sample of African countries, chosen with a view to testing his hypothesis methodically against prima facie contrary cases: he simply proposes it as a general or common pattern, whose accuracy or validity is supported only by reference to rather general examples from various countries.[28] Nor, on the other hand, does he try out any alternative hypotheses that might perhaps explain the alleged facts as well or better. In effect, he proposes as a general fact to be explained what is, in important details, not clearly shown to be true in the first place,[29] and then explains it in terms of a rational-choice model, giving illustrations that conform to the model, and not comparing its explanatory power systematically with that of any other model.

[27] Robert H. Bates, 'A Reply', *World Development* 21/6, 1993, pp. 1079–80.

[28] In his reply to Stein and Wilson Bates accepted the charge that he had relied on 'stylized facts' (something Green and Shapiro, in *Pathologies of Rational Choice Theory* (op. cit., p. 43), call 'projecting evidence from theory'), and agreed that comparative research was called for. But it does not seem from his subsequent publications that he himself is interested in undertaking this.

[29] Sara Berry, for example, questions whether the real pattern of land rights in Kenya is as Bates supposes it to be, and is doubtful how far access to land has been determined by statutory property rights in either colonial or post-colonial Kenya: i.e. both the explanandum and the explanans may be questioned ('Understanding Agricultural Policy in Africa', op. cit., p. 1059); see also her *No Condition is Permanent: The Social Dynamics of Agrarian Change in Sub-Saharan Africa* (Madison: University of Wisconsin Press, 1992), which gives strong general grounds for doubting the value of supposing that we can understand the actions of people in rural Africa by assuming that they assign priority to their immediate material interests (or indeed any other fixed set of interests), as Bates's model does. Pauline Peters argues that a comparison of a variety of societies in which there is growing population pressure on land often shows property rights in land becoming less exclusive, rather than more, contrary to what Bates assumes ('Is "Rational Choice" the Best Choice?', op. cit., p. 1070). And Manfred Bienefeld concluded from a detailed study in 1990 that in Tanzania 'the evidence does not show the existence of massive misappropriated rents or wilful efficiency losses whose recovery might allow producer prices to be systematically raised by market forces' ('Structural Adjustment and Rural Labour Markets in Tanzania', World Employment Programme Research Paper WEP 10-6/WP102, November 1990, p. 117). One is bound to wonder how far Tanzania really is exceptional, and also, to the extent that it is, how Bates would account for its exceptionalism within the framework of his model.

Bates does acknowledge some prima facie counter-evidence in the shape of 'variations' in policy among African states, which he attributes to factors specific to each country. These are 'historical factors' (a virtually limitless category, it should be noted), 'other claimants' for resources (i.e. other than farmers, such as mines or manufacturing, and government itself) and 'characteristics of production' (such as the nature of the crops or the structure of the economy). This, however, is 'domain restriction' — i.e. retaining a general explanation in face of counter-examples by allowing that it does not work in particular circumstances.[30] This may be legitimate if there are good general reasons for thinking that the theory is valid and useful, even when these domains of application are excluded. But it becomes pretty arbitrary when the restrictions are as wide as in this case. Unless we are already persuaded, on prior grounds, that the rational-choice model is too valuable to give up, we might well prefer to make historical factors, or the characteristics of production, or some other factor or set of factors, the central explanation. Yet we do not have such prior grounds; establishing grounds for a rational choice-based explanation of the African agrarian problem is precisely what Bates is supposed to be doing.

But the purpose of this discussion is not to try to duplicate the detailed critique of Bates's work that others have put forward. I am concerned, rather, with understanding the impact it has had on US political scientists concerned with development. And part of that impact is, in spite of the shortcomings that people have pointed out, the sense that Bates's work conveys of being a distinctively logical or 'rigorous' analysis. For this reason it is worth pausing, however briefly, over an example of Bates's style of argument, which, on closer inspection, displays not rigour, but precisely the kind of reasoning I have been objecting to.

Bates reasons as follows. A first hypothesis to explain why prices are set low for the mass of small farmers is that governments want to transfer surplus from farming to industry in order to enhance the national economy's long-run growth potential; this can be accomplished by obliging the farmers to sell their produce to the state at low prices and then selling it on world markets at higher prices and keeping the difference. But, with regard to food crops for *domestic* consumption, this logic would not apply. For such crops, to ensure an adequate supply, avoid food imports and benefit the mass of small farmers as well, prices should be raised; but in practice governments systematically prefer not to raise food-crop prices but to subsidize farm inputs, such as seed, fertilizer and machinery, which are primarily used by large-scale farmers. A hypothesis that fits both these facts, rather than just one, is that governments do not aim to maximize the social welfare but to serve private interests (namely, their own and those of large-scale farmers, bureaucrats, urban workers, importers and the like). Finally, a still more specific hypothesis deals with the fact that governments

[30] Green and Shapiro, *Pathologies of Rational Choice Theory*, op. cit., p. 45.

'get away' with this — i.e. the fact that the rural majority who are injured by these policies fail to get them changed — by invoking the free-rider theorem. The government responds to pressure only from those who, being few or (in the case of urban workers) concentrated, can overcome the problems of mobilizing collective political pressure, and is able to ignore the large and dispersed rural majority, who can't.

But it is not difficult to think of a combination of other hypotheses, at the same general level of abstraction, that would cover the same alleged facts. For example, the 'entrenchment' of interests around the inherited colonial structure for regulating the economy, and in particular marketing boards and import-substitution industrialization policies, would account for both the price structure and the policy instruments typically preferred for encouraging food crop production.[31] So would the no less ubiquitous pressure of foreign 'donors' on government ministries to come up with discrete projects for expanding food production, preferably involving some use of equipment supplied by firms in the 'donor' country. And the relative durability of government policies that favour large farmers and hurt small farmers (to the extent that this is true) might be explained in terms of the fact that in most African countries precapitalist relations of production still predominate, at least in form and often in substance, so that most people tend to see policies favouring notables from their own ethnic group more as favouring their ethnic group than as discriminating against small farmers.

This is not to say that these hypotheses are better than Bates's, but they do seem equally plausible. Bates, however, is committed to his hypotheses; his approach is to assert them, and to cite facts that seem to conform with them, or that are at least not inconsistent with them. And the impression of power and elegance that his argument gives seems at least partly due to the obliging way in which the evidence does fall into place when handled in this way.[32]

One more observation about the evidential base of Bates's work is prompted by his 1989 study, *Beyond the Miracle of the Market*. Instead of undertaking the kind of comparative research that was surely needed, in the mid-1980s Bates went to a single African country, Kenya, and studied the evolution of its agricultural economic policies generally, from colonial times to the present.[33] Willingness on the part of a political scientist of Bates's stature to go back and do fieldwork in a particular African country is exceptional, and yet the exercise was not calculated to deal with the crucial problem of testing the propositions of his theory. Instead, what

[31] See Gunilla Andrae and Bjorn Beckman, *The Wheat Trap* (London: Zed Press, 1985).

[32] I must admit I have been guilty of something rather similar in an early paper which argued that the number of political parties could be explained by sole reference to the electoral system, using an analogy with the 'cobweb theorem' of the pig production cycle — a youthful excess of which friends occasionally take an innocent pleasure in reminding me.

[33] R.H. Bates, *Beyond the Miracle of the Market* (Cambridge: Cambridge University Press, 1989).

resulted was a set of proposals to reinterpret what was, for the most part, a familiar story (to specialists at least), in the light of the theorems of the NPE — from the origins of Mau Mau to post-independence food crises.

The most general chapter (Chapter 3) examines 'the institutions that underlie agriculture', 'the politics of Kenya's single-party system' and 'the interaction between economic institutions and politics, arguing that the institutions put in place to enhance the productive growth of agriculture furnish as well resources which help the managers of Kenya's single national party to maintain themselves in power' (p. 73). Bates shows how anxiety that the African smallholders who were given land in the former White Highlands at independence might not repay the foreign loans with which the land had been purchased from the departing whites dictated elaborate bureaucratic controls over the marketing of their crops ('the result was an intrusive infusion of bureaucratic regulation into Kenya's rural society').[34] He suggests that the high cost of monitoring quality in smallholder production of sugar and tea led to close control over farmers' use of inputs as a cheaper surrogate for quality control. Similarly, variations in scale economies have dictated distinctive variations in the organization and production of different crops, which in turn have had effects on the distribution of income and hence on the patterns of political conflict in the respective crop zones, such as those that occurred between large- and small-scale dairy farmers in the 1970s.

Two kinds of claim are reiterated throughout the discussion: one is that we can learn more about particular aspects of Kenyan history by looking at them through the lens of the new political economy, the other is that the exercise leads to significant revisions of the theories of the NPE.[35] The second kind of claim will not be considered here, since I am concerned with the explanatory power of the theory rather than the evolution of its internal structure.[36] As to the first kind of claim, however, whether it is persuasive is perhaps to some extent a personal matter. People familiar with Kenya are

[34] A perhaps symptomatic exaggeration, given that settlement schemes accounted for only a modest portion of Kenya's rural society, and that other sectors were also bureaucratically regulated for quite different reasons.

[35] Instances of the first kind of claim abound: see, for example, Bates, *Beyond the Miracle of the Market* (op. cit.), pp. 34–37, 71, 89–91.

[36] Bates's claims to be improving institutional theory often seem to be essentially the obverse of his tendency always to look for a way of retaining his rational-choice model rather than abandon it, or severely limit its explanatory claims, in face of recalcitrant evidence. For instance, in *Beyond the Miracle* (op. cit., pp. 85–89), Bates concludes that Olson's theory of collective action needs revision. According to Olson's theory, Kenya's small cotton, sugar and tea growers should have lost out relative to the few well-organized large companies that processed their crops, but, in practice, the farmers did relatively well and the processors often made losses or even went bankrupt. Bates adumbrates a theoretical refinement to the effect that where 'the few' and 'the large' have made immobile capital investments they 'may', as a result, have no economic alternatives; as a result political power shifts to 'the many' and 'the small', who, in spite of their relative powerlessness arising from the free-rider problem, have (or 'may' have) alternatives. This illustrates the fundamental [continued on page 94]

likely to find nothing very surprising in the things Bates represents as new insights, except to feel some doubt that institutions play quite such a determinant role as Bates believes. (In particular, his effort to explain the switch in 1978 from the Kikuyu-based Kenyatta regime to the Kalenjin-based regime of Moi within the assumptions of rational-choice theory — by arguing that it represented the ultimate victory of small farmers in their capacity as voters — involves a degree of reductionism so heroic, and leaves the implications of his basic theorems so indeterminate, as to make one less rather than more inclined to follow his example.) On the other hand, few, even among specialists, will not learn something from the insights, especially about middle-level agricultural institutions, that the NPE approach offers in Bates's hands. Yet the fact remains that the study is by its nature unable to strengthen the case for accepting Bates's thesis about the determinants of African agricultural policy-making in general.

The conclusion we seem driven to, then, is that Bates's proposed explanation for crisis in African agriculture is, on closer examination, not particularly persuasive. If, therefore, the influence of his work depended on the strength of his conclusions alone, it would be hard to understand; and so we must consider other reasons as well. It is most convenient to do this after looking briefly at the work of North as well. Before doing so, however, there is one more observation that needs to be made about the third kind of objection listed above to Bates's analysis, i.e. that in spite of his method's avowed indifference to the contents of people's preferences, Bates never entertains any hypothesis other than that they put their economic interests first. Bates replied to this criticism by stating that his choice of hypothesis was determined by the things he was trying to explain, which were issues of economic policy.[37] By implication, had he been studying an issue of social or cultural policy, a different hypothesis might have been more appropriate. But this is to assume that, in relation to an issue of economic policy, it is natural that people will put economic interests first.

But what justifies this assumption? Not, anthropologists tell us, what people actually say; in their expressed views about economic matters all sorts of other motives enter in, quite as much as the desire to gain (or not to lose) income. The source of the assumption is, rather, the underlying premiss of all Western economic thought since Adam Smith that market relations already permeate and even predominate in social life.[38]

But as Karl Polanyi famously pointed out, the idea of a 'market society'

methodological problem. Any simple model can be modified in this way at the cost of loss of generality (and determinacy — those 'mays'). Faced with the choice of retaining an explanation founded on institutional logic by adding qualifications of this kind or enlarging the explanation to include other determinants of behaviour, the choice we make becomes a strategic question that cannot be settled by merely asserting the preferability of choice-theoretic explanations. For a similar treatment of counter-evidence about African governments' crop-pricing policies, see Bates, 'Agricultural Markets in Africa', in his edited volume, *Toward a Political Economy of Development*, op. cit., p. 349.

[37] Bates, 'A Reply', op. cit., p. 1079.

(i.e. a society subordinated to the market) was in truth the key idea of a novel economic doctrine (the doctrine of the 'self-regulating' market, or *laissez-faire* economics) which, taken literally, was incompatible with the survival of society itself. If land and labour become pure commodities, so that most people are driven off the land and wages are driven down to or even below subsistence, and work, housing, community and family are subordinated to the ruthless, fluctuating and accelerating disorganization imposed by market forces, life is liable soon to become physically and psychically insupportable. For this reason, Polanyi pointed out, there has always been a reaction against the market society doctrine, making the history of capitalism a 'double movement', not a single one: the drive to expand the market and liberate it from constraints has always been countered by an effort to rein it in and resubordinate it to social needs.

From this point of view, then, it is not politics that 'impair' market efficiency, but market forces that conflict with social goals; and in reality what is at stake in Africa is precisely a conflict between the principles of 'market society' and alternative conceptions — some traditional, some modern — of collective welfare. The assumption, then, that in dealing with 'economic matters' rational people act primarily for material advantage is by no means a 'natural' starting point, as Bates implied; it is a highly political one, which takes as 'natural' what is in fact at stake in the struggle for Africa's future.[39]

The basic assumptions of the NPE are thus not by any means neutral; at the very least, they need to be justified heuristically, i.e. by being shown to explain what happens better than any other assumption. But, as we have seen, the evidence offered by Bates for this is not convincing; and the fact that his work was taken up by the neo-liberals at the World Bank as evidence supporting their call for reducing the role of politics in Africa is not accidental. It flows from the fact that a materialistic conception of individualism is the starting-point of Bates's analysis, notwithstanding his declaration that his approach is neutral with respect to the values to be maximized.

Douglass North's general theory of development

North's starting-point is to put in question three central assumptions which underlie neo-classical economic thought but which manifestly have to be abandoned by anyone who wants to extend neo-classical analysis to whole economies and to long-term change: (i) the incentive structure is such that individuals can 'capture' a sufficient share of the social returns

[38] Robert Brenner, 'The Origins of Capitalist Development', *New Left Review* 104, 1977, pp. 25–92.

[39] Pauline Peters ('Is "Rational Choice" the Best Choice?', op. cit., pp. 1067–69) makes a similar complaint about Bates's concept of 'rationality', although she tends to argue more from empirical evidence that Africans actually seek to maximize non-economic values.

(i.e. benefits shared by all) of any investment the individual makes (e.g. of tenancy laws); (ii) there are positive returns to savings — this implies secure property rights; and (iii) inefficient institutions are eliminated by competition.[40] In the real world, and in relation to whole economies rather than firms, none of these may apply: there may be no secure property rights, such property rights as exist may be such as not to give individuals an incentive to invest, and inefficient institutions (inefficient in the sense of not maximizing long-run total output) can persist for long periods. North's theory of development is a theory of how these problems are overcome.

He differs from Bates in that his theory applies to the world in general and holds in the very long term. He also introduces a general theory of the state, again in the neo-classical mode. The state has a comparative advantage in violence; it trades protection and justice for revenue. It is constrained by the threat of greater violence from key 'constituents' (rivals to the throne, potential revolutionaries or makers of *coups d'état*) and from other states. To stay in power rulers need legitimacy too. The state therefore has to establish property rights which both meet its needs for revenue and satisfy the demands of its most powerful 'constituents' (subjects/citizens). The resulting property rights may or may not yield an incentive structure that is socially efficient in the above sense (e.g. the state may sell monopolies to powerful groups, which may be a cheap way to get revenue but inhibits market efficiency).

To make this model into a general theory of development, North has to account for changes over time in the institutions that determine the incentive structure. In his view, population growth obliged people to move from hunting and gathering to settled agriculture, and this produced the need to defend territory, and hence states. Thereafter change was driven by population growth plus further changes in technology, especially military technology, which forced states to make changes that would raise total output so that they got enough revenue to pay for the new military technology needed for defence. The changes they made included administrative centralization and territorial expansion to achieve the necessary scale economies.

But, since states, economic systems and social systems constitute complex and closely interlocked wholes, change is hard to achieve, and, typically, more or less inefficient economic systems persist for long periods, following paths determined by earlier institutional choices ('path dependency' — a general concept, drawn from examples such as the narrow-gauge railway tracks first adopted in the UK, or the 'qwerty' layout on the first typewriter keyboard, i.e. cases where the initial choice of a particular technology 'locks' an economy into a sequence of adaptive investments that block the later adoption of more efficient technologies). What, then, does bring about more efficient institutions, short of catastrophe (e.g. being conquered by another state)? Within the assumptions of rational-choice

[40] North, *Structure and Change in Economic History*, op. cit., pp. 5–7.

theory the free-rider problem makes it unlikely that social movements will bring about any change, let alone a fundamental, revolutionary one; so change should be more likely to come from rulers, who don't have this problem (they are one person, or a few). But historically many changes have come from large-scale social movements. Therefore, North says, a 'positive theory of ideology' is needed to supplement the axioms of rational choice and account for how the free-rider problem is overcome to bring about major institutional changes.[41]

Pending the creation of such a theory, North reasons that most change is actually incremental, arising as a by-product of multiple adjustments to changes in prices. Radical changes will typically be diluted in practice, too, by the influence of the myriad informal institutional rules embodied in culture and in social relations of all kinds. North's general idea is that some of the organizations that are formed to take advantage of the opportunities provided by the existing institutional structure come to want changes in the structure; and, while the greater likelihood is that the result will tend to reinforce the existing 'path' which so much is 'locked into', it may none the less alter it. Whether it does or not depends on the organizations and the institutions in question, and the strategic possibilities the mix gives rise to.

This a very abstract set of formulations, to say the least; and the summary I have given certainly does not convey the persuasive flavour or the historical sweep of North's writing (especially in his 1981 book; by 1990 he seems to be spinning his wheels a bit).[42] But I do not think that it fundamentally misrepresents its essential character, namely that it is a set of general concepts for interpreting history, not a substantive theory. North makes very few positive claims, and virtually none that concern the contemporary world; most of his illustrations of the theory as a theory of development come from periods before the prevalence of industrial capitalism.[43] And this raises the question: what is the relation between North's work and Marx's?

In brief, the answer is that North sees Marx's theory as a special case of the kind of general theory he is seeking to formulate. His references to

[41] 'Without an explicit theory of ideology or, more generally, of the sociology of knowledge there are immense gaps in our ability to account for either the current allocation of resources or historical change. In addition to being unable to resolve the fundamental dilemma of the free rider problem we cannot explain the enormous investment that every society makes in legitimacy...Neoclassical economics...may account very well for the performance of an economy at a moment of time...but it does not and cannot explain the dynamics of social change' (ibid., pp. 47 and 57).

[42] Douglas C. North, *Institutions, Institutional Change and Economic Performance* (Cambridge: Cambridge University Press, 1990).

[43] The chief exception to this is his discussion of American land tenure law and the socio-economic benefits that flowed from it. The serious analysis — i.e. serious in the sense of aiming at practical conclusions on which policy might be based — of different national forms of industrial capitalism, such as those put forward by historically orientated comparativists like Peter Hall, Michel Albert or Lester Thurow, could clearly not be[continued on page 98]

Marx are sympathetic, but always end with a comment to the effect that Marx has been proved wrong in this or that respect — that socialism did not prove to be more efficient, that human nature did not change under socialism, and so on. However, what is clear is that, if Marx made errors, it was because he put forward a substantive theory of institutional change capable of being wrong, which North does not. If one doesn't accept Marx's theory, the point must be to improve on it through criticism and modification, or total replacement. But this cannot be done within the limits of rational-choice theory; this is what North's call for a positive theory of ideology really acknowledges. What sounds like a call to decisive theoretical action is actually an admission of theoretical impotence.[44]

This judgement might seem to be contradicted by the fact that in his *Structure and Change* North offers a rational-choice-based interpretation of world economic history from palaeolithic times to the present, and, breathtaking as this seems, it is the most attractive part of the book. North argues that in any explanation of the rise in productivity that has in fact occurred in history property rights enforced by states must be central. This yields a series of observations of the following type (here North is referring to the consequences of the spread of settled agriculture):

> There can be no doubt that substantial growth did occur...I refer to per capita growth in income that meant that for lengthy intervals output grew faster than population...We do not have to search far to account for the economic sources of the productivity increase that underlay the thriving civilizations of the ancient world. The improving institutional organization of the state that we observe in the sequence from Egypt to Persia to Greece to Rome led to the reduction in transaction costs, growing regional specialization, and widening of the markets... In addition...there was immense technological change over these eight millennia [down to the second century AD].[45]

Can anyone doubt that this must be true? And when the fall of Rome is explained by the same kind of reasoning (as North says, 'the economies of scale that made possible a single empire governing the entire Mediterranean world had disappeared')[46] — and then the rise and the decline of feudalism, the emergence of England as the seed-bed of industrialization, and the dynamics of the industrial revolution itself — it is hard not to be impressed. North's evolutionary reading of world history seems more

based on such narrow foundations (see Peter A. Hall, *Governing the Economy: The Politics of State Intervention in Britain and France* (New York: Oxford University Press, 1986); Michel Albert, *Capitalism Against Capitalism* (London: Whurr, 1993); and Lester Thurow, *Head to Head* (London: Nicholas Brealey, 1994)).

[44] See John Toye, 'The New Institutional Economics and Its Implications for Development Theory', forthcoming in Harriss *et al.* (eds), *The New Institutional Economics and Third World Development,* op. cit., draft p. 22: 'The main weakness of the NIE as a grand theory of socio-economic development is that it is empty.' I am grateful to Professor Toye for an opportunity to read a draft version of this paper.

[45] *Structure and Change*, pp. 110–11.

[46] Ibid., p. 123.

plausible, too — if less poetic — than Marx's architectonic conception of epochs initiated and terminated by decisive struggles between oppressing and oppressed classes.[47]

But, when the spell is broken, it can't be denied that North does with the history of the world what Bates does with the history of Kenya: it is really a series of illustrations of the alleged explanatory power of institutions, and especially of property rights. That these are important seems likely; but what no overview of this kind can show is how important, or in what particular ways and contexts, which is what we really need to know. It is a matter of speculation. It seems unlikely that property rights are sufficient to ensure growth; perhaps they are necessary, but, in the absence of a theory of what determines institutions, we cannot even be sure of this. North's account of world history may have its charms but it does not strengthen, let alone confirm, the claims of the NPE to be a general theory of development.[48]

What is the appeal of the NPE?

If the NPE rests on inconsistent axioms and, as a general theory of development, is both implausible and empirically unsubstantiated, what accounts for its influence among American political scientists in this field? Part of the answer is no doubt the real contributions the 'new institutionalists' have made to middle-level institutional and policy analysis. But I think there are two other reasons worth focusing on as well.[49]

One is the persistent desire, especially among US political scientists working in the shadow of neo-classical economists, for a powerful, general, hypothetico-deductive theory. The appeal of any such theory is, obviously, that, if the real world seems to conform to it, its very simplicity becomes its chief merit; i.e. it then appears to have identified causal relations so

[47] Anyone who finds talk about capitalism and socialism uncomfortable or *passé* will also welcome North's conclusion that it is not the struggle between capital and labour that is central to the fate of modern society (Marx and Polanyi, he assures us, both got this wrong), but the search for a balance between the social benefits of specialization and scale economies, and the returns to using political power to reduce the social costs (see *Structure and Change*, op. cit., especially pp. 180–86).

[48] One of the attractive aspects of Bates's work is his openness to criticism, an example of which is the inclusion in his edited collection, *Towards a Political Economy of Development* (op. cit.) of a tough-minded critique of North's theory (and the somewhat related theories of Mancur Olson and Robert Gilpin), which concurs with this conclusion, namely Ronald Rogowski, 'Structure, Growth and Power: Three Rationalist Accounts', pp. 300–29.

[49] A third reason must be the aggressive drive of many exponents of rational-choice-based approaches for control over appointments, promotions and the direction of research students in many departments of political science in the USA (and at least one in Canada). It is reminiscent of the attempts previously made by behaviouralists and, in political philosophy, 'Straussians' to enforce their respective 'paradigms', and the phenomenon no doubt signifies something about the scientific nature of these successive schools.

powerful that all other aspects of reality, however connected they seem, can be safely disregarded. Even when the real world does not conform, we can still feel a gain: we can either rule out the causal relations suggested by the model, or we can go back to the model and adjust it, and see if the fit is improved.

But in the complex, non-repetitive, 'agonistic' field of politics, what we typically get from such theories is at most flashes of illumination, which are immediately overlaid by counter-evidence that forces us to acknowledge that, for the relationships postulated by the model to be preponderantly, let alone exclusively, determining, the world would have to behave quite differently from the way it actually does. Normally, that is, we can at most use this method like a kind of probe, a way 'into' the complexity of the causal systems that are at work. There is nothing wrong with pushing the probe as far as it will go in as many places as seem useful; and it is a useful antidote to forms of analysis that are essentially speculative, unwilling to clarify their sometimes obscure assertions by translating them into language that would make them susceptible of being evaluated by model construction and testing.[50]

What is dangerous, though, as Przeworski says, is to start believing your models, as both Bates and North do not always avoid doing.[51] Perhaps it is an occupational hazard of the hypothetico-deductive mind, which may be especially reluctant to accept the limited degree to which the social world is ultimately ordered at all; but the 'Cartesian' wish to find firm foundations on which to base our understanding is very general (post-structuralism may have discredited 'foundationalism' in the sense of a belief that firm foundations for statements about the world can be found by close inspection of that world itself, but we seem addicted to it). So, when Bates says that he is concerned to find the 'microfoundations' of large-scale social phenomena or 'to cut through the emotion and turmoil of the events surrounding Mau Mau and to highlight the *sparse underlying structure* that generated the grievances fueling the revolt', we are predisposed to believe that this is what his method has, indeed, accomplished.[52] To get such reassuring

[50] Some Marxist analysis has been guilty of the latter fault, whether from a determination at all costs to keep the 'totality' in view or from fidelity to dogma. To the extent that the project of the 'analytical' or 'rational-choice' Marxists is to break with this weakness in the Marxist tradition (as is the case, at least, with Jon Elster's *Making Sense of Marx*, op. cit.), I agree with it.

[51] See note 26 above. Bates seems ambivalent about this, as may be seen by contrasting his statement, in his 'Reply' to his critics (op. cit., p. 1081), that he will not lament the demise of rational-choice theory if better approaches 'come along', with his constant, and I think symptomatic, insistence throughout his 1987 study of Kenya, in particular, that the NPE approach contains 'lessons' of all kinds, 'dramatically confirms' this or that, and so on.

[52] See Bates, *Towards a Political Economy of Development*, op. cit., p. 6, and *Beyond the Miracle of the Market*, op. cit., p. 12 (my emphasis). Note how the positive appeal of parsimony (sparseness) is enlisted to support the claim to have uncovered what generated the 'grievances fueling the revolt' — all of them? Is it really plausible that people would have taken the

knowledge we are willing, at least momentarily, to suspend our common-sense awareness of the vast range of other determinants that are actually involved, and to accept rather uncritically the account given of both the alleged facts to be explained and those that are said to explain them.

Another reason for the appeal of the NPE may be its very reductionism. Its social ontology of individuals, groups, goods and costs is indeed so 'sparse' that it is also politically anaesthetized. For example, here are some of the claims made by Nabli and Nugent for the new institutional economics (NIE) as a contribution to the theory of development:

> Institutional change can be considered at the heart of the long-run process of economic development, providing the missing link between development and growth... Indeed, it might be appropriate to define economic development as economic growth accompanied by 'efficient' institutional change... [the preceding discussion of the NIE shows that] by affecting transaction costs and coordination possibilities, institutions can have the effect of either facilitating or retarding economic growth... On the other hand... running through all the principal themes of the NIE is the notion that economic growth can have profound impacts on institutions. Economic growth may well induce changes in contractual choices, the relative importance and character of markets, the extent of private property rights, the relative importance of different constraints, the relative position and power of interest groups and organizations... and the degree of internationalization of the economy... Economic growth can render some existing institutions, such as norms, rules and policies, redundant, and require others to come into existence.[53]

To whom, it must be asked, might this distinctly bland formulation seem more appealing than the following more famous one, which clearly covers the same ground?

> In the social production of their life, men enter into definite relations that are indispensable and independent of their will, relations of production which correspond to a definite stage of development of their material productive forces. The sum total of these relations of production constitutes the economic structure of society, the real foundation, on which rises a legal and political superstructure... At a certain stage of their development, the material productive forces of society come in conflict with... the property relations within which they have been at work hitherto... [etc.]?[54]

Is it far-fetched to imagine that part of the appeal of the NPE lies in its

terrible oaths, and accepted the risks and the suffering of such a grossly unequal war, without having undergone a religious, national, ethnic and racial mobilization over several generations, a mobilization whose causes were necessarily far too complex to be accounted for by any 'sparse' structure?

[53] Page 1342 of Mustapha K. Nabli and Jeffrey B. Nugent, 'The New Institutional Economics and Its Applicability to Development', *World Development* 17/9, pp. 1333–47).

[54] Karl Marx, 'Preface' to *A Contribution to a Critique of Political Economy* (1859), in Robert Tucker (ed.), *The Marx–Engels Reader* (New York: Norton, 1978), p. 4.

very blandness and indeterminacy, its essential lack of substance? And in the fact that it appears to offer a way of talking about the same problems that Marxism tackles, but in a way that makes Marxism only a 'special case' (and a faulty one at that) of a (proposed) much more general theory? And, if the price of this greater generality is that it can only speak of what 'can' or 'may well' happen (except for that inadvertent 'require' — wherever did that come from?), until the new, improved general theory has indeed been constructed that will actually suggest what must happen and why, is this, perhaps, a price that many people — especially in the US context — will be quite content to pay?

For, as Mahmood Mamdani has pointed out in relation to Africa, First World scholars working on underdeveloped countries are always constrained by their dependence on state or private sources of funding, which 'discipline and shape the very process of the production of knowledge'. And they are often also unconnected to popular struggles either in the Third World or in their own country.

> Where not anchored in popular struggles, whether in Africa or at home, the Africanist tends to succumb to the pressure of institutional ideologies... [this] tendency...flows from the 'realism' the Africanist embraces, whether in succumbing to these constraints, or in an isolated rearguard struggle against them... Whatever his or her methodological preference, the analysis tends to take on a character either static and descriptive, or speculative and scholastic...From the time our scholar becomes 'realistic', he or she begins to take the 'official' standpoint as the point of departure in the process of the production of knowledge. The tendency is for the 'realistic' point of view to overlap with the 'official' point of view. The Africanist begins to speak the language of a consultant.[55]

It is precisely to this situation — and especially in the USA where, in ground already prepared by a tradition of intense intolerance towards radical politics and left-wing thinkers, neo-liberalism assumes its most dogmatic and aggressive form — that the NPE seems to have something particularly valuable to offer (in addition to the appeal of its seeming rigour, esoteric technical knowledge, and so on): namely, a way of talking about some of the core issues dealt with in Marxist theory, without the professional penalties.

To take one more example. Marx theorized that at a certain stage of the development of the forces of production, the existing property relations cease to be 'forms of development' of the productive forces and turn into their 'fetters'. Marx's idea of 'fettering' has obvious similarities with the concept of 'path dependence' that plays such a central role in the NPE

[55] Page 7 of Mahmood Mamdani, 'A Glimpse at African Studies, Made in USA', *CODESRIA Bulletin* 2, 1990, pp. 7–11. Mamdani's comments were prompted by the Proceedings of the Inaugural Seminar of the Governance in Africa Program of the Carter Center of Emory University, 17–18 February 1989, subsequently published as *Beyond Autocracy in Africa* (Atlanta: Carter Center of Emory University, 1989). I am grateful to Dr Agnes Callamard for drawing this article to my attention.

when considered as a theory of development;[56] the essential difference is that, in the literature on 'path dependence', the dependency is purely technological, not technological and political, and the starting-point of the path always appears as accidental, not class-determined. The NPE thus seems to allow one to discuss the phenomenon of something like 'fettering' with reference to no more alarming an example than the layout of the typewriter keyboard.

The 'rational-choice Marxist' Jon Elster has suggested that Marx's theory of history can be read as an exploration of the problem of unanticipated consequences — of why people's individual choices lead to results very different from those they hope for;[57] to the extent that one accepts such a reading of Marx, it is another way of stating the seeming convergence between his project and that of the NPE. But because the NPE is so abstract it is as compatible with neo-liberalism as with Marxism — and all points in between (Bates, for example, is also probably quite right in suggesting that 'a new generation, seeking middle ground between the champions of the market and the defenders of the state, finds in the new institutionalism a justification for basing development efforts on community action and civic engagement').[58]

Perhaps I have exaggerated the influence of the NPE. I hope so; its ambitions are not matched by its potential. Meanwhile, it remains one of the bitterer ironies of the last two decades that so many of the poorest people in the world were forced to become even poorer, and in many cases to endure unspeakable suffering, through the imposition of Structural Adjustment Programmes justified in terms of 'rational choice'.

[56] See, for example, pp. 1389–90 of Pranab Bardhan, 'The New Institutional Economics and Development Theory: A Brief Critical Assessment', *World Development* 17/9, 1989, pp. 1389–95, and Bates, 'A Reply', op. cit., p. 1078.

[57] In Elster, *Making Sense of Marx*, op. cit.

[58] Bates, 'Social Dilemmas and Rational Individuals', op. cit., p. 2. The NPE's discourse of 'choice' also seems to have influenced some African intellectuals, who, in an ironic twist of the slogan that Africans must be seen not as victims but as subjects of their own history, are often surprisingly ready to attribute the continent's situation to choices that 'Africans themselves' have made; could the discourse of 'choice' popularized by the NPE have something to do with this? A related discourse that seems relevant here is that of the 'informal' sector. The leading popular text for this discourse, Hernando de Soto's *The Other Path* (New York: Harper and Row, 1989), draws on 'transaction cost' analysis to argue for 'another path' to development through eliminating 'mercantilist' and 'redistributive' state interventions in the market that thwart the entrepreneurial potential of the 'little people'; de Soto thus links a populist discourse of 'the marginalized subject' to neo-liberal market ideology (the book comes with an endorsement from former President George Bush on the back cover).

II

Development Theory & Africa

5

African Economic Development in Theory & Practice

The character of African economic development in the twenty years 1960 to 1980 is strangely obscure. This has a good deal to do with the size and diversity of the continent — Nigeria booming, Ethiopia starving, and so on — but it is due even more to the filtration of the African experience through successive layers of interpretative theory. One can see this from the sort of questions people ask: Have the African economies become less or more 'dependent' or 'externally orientated'? How far is their new technology 'appropriate'? Are they less vulnerable than formerly to 'unequal exchange'? How far have 'basic needs' been met? These are not primarily factual questions. They testify to the extraordinary degree to which African reality has been overlaid by theory. It is as if the poverty of so many millions in Africa were too painful to contemplate without being immediately set in the context of a theory that can somehow explain it, and at the same time show the way to ending it. The motive is good, but the pitfalls are many. Today, the prevailing view of African economic development is pessimistic — profoundly disappointed at the post-independence record and gloomy about future prospects. This view, however, is a reaction to a previous moment of exaggerated optimism. Neither really helps us to understand what is happening in Africa. To be able to come a little closer to that reality, we need to stand back somewhat from the interplay of hopes and fears and theories that reflect and may so easily amplify them.

The phase of optimism

It is easy to forget just how unqualified was the optimism about African development that reigned in the early 1960s, just as the formal decolonization of Africa was being completed. Ten newly independent African countries were among the eighteen United Nations (UN) delegations that

* Reprinted by permission of *Daedalus*, Journal of the American Academy of Arts and Sciences, from the issue entitled 'Black Africa: A Generation After Independence', Spring 1982, Volume 111, Number 2.

persuaded the General Assembly to declare the goal of the UN Development Decade to be 'the attainment in each less developed country of a substantial increase in the rate of growth, with each country setting its own target, taking as the objective a *minimum* annual rate of growth of aggregate national income of 5 percent at the end of the decade'.[1] For the Second Development Decade (1970–80), the target was raised to 6% and, in both decades, ambitious targets for industrialization were also set. These targets were far higher than the historical achievement of most of the industrialized countries; Britain's growth rate, for example, never reached a 2% average from 1853 to 1913, and even the US and German rates of growth of gross national product (GNP) down to the Second World War were well below 5% a year on average.

What accounted for this optimism? Partly it was a result of the general optimism caused by post-war economic recovery, the unprecedented growth rates of Germany, France and Japan, and the belief that in Keynesian demand management, combined with sound international monetary arrangements, a technique had at last been found that could maintain a permanently rising volume of international trade. There was also the example of the Soviet Union's spectacular growth between the wars, based on five-year plans, a technique also adopted by India after its independence. And there were the African leaders' keen sense of the waste of their peoples' human potential under colonialism and the ambition to catch up with the industrial countries, by taking advantage of their technology and of the aid that was expected to flow from them in growing amounts.

The results of the two 'development decades' in Africa certainly belied this optimism. More than four-fifths of the sub-Saharan African countries still fell in the low-income category of developing countries (with annual per capita incomes of less than US$360 in 1978), and their average rate of growth per capita of 0.9% over the two decades was the lowest of all the regions of the Third World.[2] Worse, the World Bank's most optimistic forecast was that the African countries would experience only a 1% average annual per capita growth rate in the 1980s; more probably, they would on balance experience a slight decline, with a growing number of people in 'absolute poverty'.[3] It is true that gross rates of growth (undivided by population growth) were on average quite strong (although not up to the 'development decade' targets): for the low-income African countries, the average annual growth of gross domestic product (GDP) was 4% in the 1960s and 2.4% from 1970 to 1978. For the eight African middle-income countries (above US$360 per capita GNP), the figures were 4.4% and 3.7%.[4] Given

[1] *Year Book of the United Nations*, 1961, p. 229, my emphasis.

[2] *World Development Report* (Washington, DC: World Bank, 1980), p. 85.

[3] Ibid., p. 12.

[4] Unweighted averages, ibid., p. 112. The middle-income sub-Saharan African countries are: Ghana (per capita GNP US$390 in 1978), Cameroon (US$460), Liberia (US$460), Zambia (US$480), Zimbabwe (US$480), People's Republic of the Congo (US$540), Nigeria (US$560), and the Côte d'Ivoire (US$840). It should be added that, of course, growth

the world recessions of the 1970s, these figures were not unimpressive; and there were great individual variations, with some African countries — not all of them oil exporters — enjoying much more rapid rates of growth. But there were also many countries that stagnated, and others that regressed disastrously — for instance, Angola and Uganda. In any case, given annual population growth rates closer to 3% than to 2% in most African countries, economic growth rates needed to be very high, by world standards, if Africans were to become better off. Moreover, statistical averages conceal severe inequalities, as seen in Table 5.1.

Table 5.1 Inequality in Africa

	GNP per capita	Proportion of GNP received by		
		Lowest 40%	Middle 40%	Top 20%
Tanzania (1967)	89	13	26	61
Uganda (1970)	126	17	36	47
Kenya (1969)	136	10	22	68
Sierra Leone (1968)	159	10	22	68
Côte d'Ivoire (1970)	247	11	32	57
Zimbabwe (1968)	252	8	23	69
Gabon (1968)	497	9	24	67

Source: M.S. Ahluwalia, 'Income Inequality', in H. Chenery *et al.*, *Redistribution with Growth* (London: Oxford University Press, 1974), p. 12.

Food production per capita had declined, and in the Sahel region, stricken by war and drought, aid agencies estimated that a million people would die of hunger and malnutrition in 1980 alone. In the meantime, development aid declined steadily from the levels of the 1960s — especially US aid, which fell from 0.53% of US GNP in 1960 to 0.18% in 1980, compared with the Second Development Decade target of 0.7%. At the same time, African indebtedness rose. By 1990, the World Bank predicted, repaying and servicing old debt would use up over 80% of all new borrowing by the African countries. Grim scenarios prevailed for the future of African resource depletion, pollution, regional co-operation, and the like.[5]

rates are often poor indicators of changes in living standards. Even if we neglect the question of income distribution (which may become more unequal with growth of the GNP), a growth of GNP may have no effect, or an adverse effect, on domestic living standards, as has often been noted, for example, in the case of growth based on mineral exports or tourism. It is also necessary to recognize the dubious basis of many national income statistics. They are cited here because no better indicator of the development of the forces of production is available at present.

[5] T. Shaw and M. Grieve, 'The Political Economy of Resources: Africa's Future in the Global Environment', *Journal of Modern African Studies* 16/1, 1978, pp. 1–32.

The pessimistic reaction — from 'modernization' to 'dependency' and 'basic needs'

Instead of leading to a systematic theoretical critique of the basis for the earlier overoptimism, the gap between those expectations and experience led to an equal and opposite tendency to pessimism. The optimism had been rationalized by theories of 'modernization'. The pessimistic reaction was rationalized in terms of 'underdevelopment' and 'dependency' and given a practical expression in poverty-orientated development programmes. It is necessary to place these successive theoretical systems in perspective.

Modernization theory was the outgrowth of a gradually expanding perception of the obstacles in the way of rapid growth in the ex-colonies. At the end of the Second World War the emphasis was heavily, if not exclusively, on the shortage of capital. By the mid-1950s, it was recognized that the problem of capital absorption — that is, effective use of capital — was, if anything, more serious than the problem of capital supply. This view lay behind the increasing emphasis on development planning in the late 1950s. Experience with planning soon revealed, however, that implementing plans was more difficult than drawing them up. This led to increased concern with development administration, administrative and management training, and even to experiments with 'infecting' Third World businessmen with 'achievement motivation'.[6] By the mid-1960s, it was apparent that none of these sectors could be isolated from the total social milieu, and there developed a concern with altering traditional values, attitudes and social practices. Modernization theory denoted the sum of these perspectives. The mid-1960s were its high point. Social scientists working in Africa were among the most influential exponents of the modernization approach, according to which development consisted of a complex transition from traditional primordial society, based on multiplex, affective and ascriptive relationships, to modern society, based on role separation, rational relations and achieved statuses. At a certain stage in this transition, a 'take-off' into sustained economic growth would become possible.

The main tenets of economic theory that belonged to this general outlook have been summarized by Killick:

- Economic development is a discontinuous process of structural transformation.
- National poverty is self-perpetuating, with low-income countries caught in a vicious circle of poverty. But, with the savings ratio as a rising function of income, growth tends to become self-sustaining above a certain critical level of income per capita.
- A 'big push' or 'critical minimum effort' is required to break out of the

[6] Some American foundation money was spent in Africa, although more in the Middle East and South East Asia, on training programmes founded on the quaintly fantastic ideas of D. McClelland, as developed in *The Achieving Society* (New York: Van Nostrand, 1961).

'low-level equilibrium trap' and achieve self-sustaining growth.

- While the 'big push' requires many inputs, its single most important ingredient is a massive increase in the ratio of investment to national income.
- Development entails industrialization which, by choice or necessity, will concentrate on satisfying the home market for manufactures by substituting for imports.

Killick adds:

> With the probable exception of [the first] item, not one of these propositions would today command general assent among Western-based or trained economists.[7]

This tells us that these theories were wrong, but not why. Their fundamental weakness was that they rested on certain illusions that modernization theory made into an explicit doctrine. They assumed that the 'backwardness' of the Third World was an 'original' backwardness, a primeval backwardness that had once been universal and could be overcome by the transmission of capital and know-how from the industrial West. But this was profoundly misleading. The backwardness of Africa was a new form of backwardness, the product of colonialism. The technological backwardness of African agriculture, for example, had been shaped and even developed by colonialism. In the mid-nineteenth century, for instance, before colonial rule, H. Kjekshus estimates that the interior of what is now mainland Tanzania carried at least 4.5 million cattle; with regard to the coastal region, he cites a German officer writing in 1891 that 'the still widespread opinion in Germany that you need only step outside of your door to shoot an antelope or a pheasant anywhere in Africa is based on a complete misunderstanding. The entire coast is as poor in game as can possibly be imagined.' It was poor in game because it was supporting a rich agricultural and pastoral economy. Kjekshus adds, 'Similar reports abound from areas that are today the country's major game sanctuaries' — areas now infested with the tsetse fly, where no cattle can live.[8] It was only after the introduction, under colonization, of rinderpest, smallpox, jigger fleas and other scourges, followed by the retreat of people, the advance of the tsetse and the forced development of export crops in the remaining areas of cultivable land, that Tanzania's agricultural backwardness acquired its modern form. And what development was achieved was done with virtually no new technical or capital inputs. Walter Rodney's comment that 'the vast majority of Africans went into colonialism with a hoe and came out with a hoe' contains a very large element of truth. Similar observations could be made for African trade, African manufacture and African administration in

[7] T. Killick, 'Trends in Development Economics and Their Relevance to Africa', *Journal of Modern African Studies* 18/2, 1980, pp. 368–69.

[8] H. Kjekshus, *Ecology Control and Economic Development in East African History* (London: Heinemann, 1977, p. 72; London: James Currey, 1995).

many parts of the continent. This is not to say that colonialism cut off a development that would otherwise have made Africa into a region of advanced industrial societies. It is only to say that, to an embarrassing degree, not only did modernization theory fail to see that African backwardness was shaped by colonialism, but it also failed to see how far the post-independence pattern of trade and investment, the patterns of aid given to local 'élites,' or the transfer of Western tastes reinforced the backward, inegalitarian structures of the ex-colonial African economies.

In Africa, the theoretical reaction to modernization, strongly inspired by Latin American 'dependency' writers, began in the late 1960s, particularly after the Arusha Declaration of 1967 in Tanzania. The essence of this reaction was to shift from ignoring many of the external determinants of African development to seeing them as primary, and as almost wholly negative, inspired as they were by the interests of foreign capital and foreign states, not the interests of the population concerned.[9] In this view, the result was an 'external orientation' of the African economies (geared to exporting primary commodities of low value and importing manufactures of high value); a process of 'unequal exchange', which limited the development of 'internal linkages'; and severe inequalities of income and wealth, which further limited the development of the domestic market, reinforced the political power of comprador ruling classes and gave rise to chronic corruption, political instability, military coups and so forth. And this process was aggravated, not relieved, by industrialization undertaken by multinational corporations, leading, among other things, to high rates of surplus transfer, to the rising use of imported inputs and to capital-intensive industrial technology.

Out of the critique of modernization theory and the apparent ineffectiveness of many of the economic development policies associated with it, there emerged the so-called poverty-orientated, or 'basic needs' approach to development. Developed largely within the International Labour Organization (ILO) and the World Bank, it was strongly influenced by a group of thinkers especially concerned with Africa (notably Hans Singer, Dudley Seers, Richard Jolly and others), and popularized by Robert McNamara during his middle years as the Bank's president. Behind this line of thought lay an acceptance, although usually tacit, of much of the dependency viewpoint. The approach sought to raise the living standards of the poorest inhabitants of Third World countries by means of policy packages — rural development, aid to the so-called informal sector, and so on — that would directly benefit them, without challenging the basic structures of the economy or the balance of political power (although it was sometimes implied that the relief of poverty would lead to both of these results), either because these were seen as unalterable, or because the proponents of such

[9] For example, the works of Samir Amin, Roger Genoud, G. Arrighi, C. Meillassoux, J.S. Saul, E.A. Brett, Walter Rodney, M. Mamdani, I.G. Shivji, J. Rweyemamu, P.-P. Rey, M. Cowen, G. Williams and R. Howard.

an approach were not enthusiastic about changing them. These thinkers were, however, distressed by the human costs of the present situation. 'Redistribution with growth' (as an early version of this line of thought was aptly titled) represented the unhappy consciousness of international development orthodoxy. It is interesting to contrast the standpoint of this approach with the views of Raul Prebisch after more than half a century of work on the development of Latin America: 'You cannot produce...fundamental transformation without changing the power structure. You cannot take the surplus from the private hands that have appropriated it without subduing their power...In my view the fundamental thing is to change the power structure in the country.'[10]

The dependency theorists and their epigones undoubtedly identified many of the forces that explain the pattern of modern Africa's development. Case studies in virtually every African country testify to the reality of the structures to which dependency theory points, and the basic needs approach, which is so deeply indebted to dependency theory, became a new orthodoxy. The late 1970s view of African development was overwhelmingly coloured by these orientations. Yet there is little doubt that they too obscure as much as they reveal. To read the African record, it is necessary to remove this theoretical envelope in its turn.

The most important shortcoming of dependency theory is that it implies that there is an alternative, and preferable, kind of development of which the dependent economies are capable, but which their dependency prevents them from achieving — when this alternative does not in fact exist as an available historical option. The core meaning of 'dependent development' or 'underdevelopment' is that 'the economies of one group of countries are conditioned by the development and expansion of others' in such a way that the development of the former is blocked.[11] But this blockage is not held to be absolute; after all, there has been some growth in almost all of the Third World countries, including the African countries. So what is said to be thwarted is some supposed alternative course of development, which would be followed if the countries were not dependent. This, however, is never specified — and for a good reason. Either what is implied is a superior (autonomous, inward-orientated) kind of capitalist development; for this to be helpful it would then have to be shown how an autonomous capitalist development could be expected to unfold without the inequalities and unevenness, the instability, crises, unemployment and wars that have characterized early capitalism elsewhere, and to do it much faster than has ever been achieved before. Or the implied suppressed alternative is a socialist path of development: in this case, it would be necessary to show the social and political forces capable of carrying through such a strategy,

[10] Interview with E. Crawley, *South*, January–February 1981, p. 32.

[11] The quotation is from T. Dos Santos, 'The Crisis of Development Theory and the Problem of Dependence in Latin America', in H. Bernstein (ed.) *Underdevelopment and Development* (Harmondsworth: Penguin, 1973), p. 76.

and that it could reasonably be expected to be superior. In most African countries, this would be difficult, to say the least. There has to be something wrong with a conceptualization the import of which is that the world should be other than it can be. This manifests itself in a radical ambiguity. Concerning Niger, for example, R. Higgot writes:

> groundnut production accounted for 65 per cent of all Niger's very feeble external revenue during the first decade of independence. As of 1976, however, uranium was accounting for the same percentage but of a larger absolute volume...While this revenue bonanza makes Niger more solvent, and able to overcome some of the distress of the drought period of the early 1970s, it also ties it much more firmly into a dependency situation within the world economy.[12]

One sees what the author is getting at. But, given that he does not see any alternative way of producing the uranium ore, or any alternative means of starting to raise the level of Niger's productive forces, is Niger's dependency on the French Atomic Energy Commission, the Pechiney Mokta group and others (which also became to some extent dependent on the government of Niger) a good thing or a bad thing? What is gained by focusing the discussion around the concept of dependency?

Dependency theorists have performed an important task of demystification. It was important to insist that all development has a definite historical and social character — that there is no such thing as development *sans phrase*, and that the development experienced in Africa is capitalist. It is important to avoid romanticizing a manufacturing development that consists of spending large sums on advertising to persuade Africans to eat Weetabix, costing twenty times as much and being one twentieth as nourishing as the maize meal they used to eat.[13] It is necessary to be alert to the multiple ways in which 'periphery capitalism' involves economic, social and political costs for periphery populations — the loss of surplus through unequal exchange in monopolistic markets or through transfer pricing, the hidden costs of 'aid', the suborning of national leaders by foreign capital, and so on. But it is another matter to elevate a catalogue of periphery capitalism's many shortcomings into a doctrine of its inability to develop the periphery — relative to an implied alternative way of going about it, which is not in fact available.

This is not to say that there are no other options. The point is rather that dependency theory implies that there is some option, superior to what at present exists, when this is often not the case. Because of this, it fails to recognize that some of what is happening under dependent development is, after all, still development: painful, wasteful and ruthless, like early capitalism everywhere, but development none the less. In some areas, exchange

[12] R. Higgot, 'Structural Dependence and Decolonisation in a West African Landlocked State: Niger', *Review of African Political Economy* 17, January–April 1980, p. 57.

[13] R. Kaplinsky, 'Inappropriate Products and Techniques: Breakfast Food in Kenya', *Review of African Political Economy* 14, January–April 1979, pp. 90–96.

has been more or less generalized, production has been at least partly reorganized and wage-labour has been increasingly detached from the land. In the sufferings of the masses — not just from famines and wars, but also from forced migration, cultural deprivation, humiliation, extortion and insecurity — there is also a certain potential for advance. The failure to recognize and grapple with this renders dependency theory misleading and hence impotent in relation to those areas where the advance has occurred.

Three other shortcomings of the dependency approach also deserve emphasis. First, like any structuralist theory, dependency theory has tended to be historically weak. It is true that dependency theorists correctly reproached modernization theorists with being ahistorical, but the 'history' to which the *dependencistas* themselves appeal is typically very linear and general, not to say universal. It presents the influence of the industrialized countries on the tropical world as a broad and continuous process of surplus extraction that has merely taken successive forms — from primitive accumulation through colonial monopolies, to sophisticated practices such as transfer pricing and monopoly pricing for technology transfers. In this perspective, little distinction is made between the different stages of development of capitalist countries at the 'centre', and the effects of these differences on relations with the periphery; or between the different resource endowments, scale, geopolitical significance and, above all, pre-existing social structures of the various periphery countries. Today it seems clear that every country must be understood in the uniqueness of its own historical development and its own distinctive relations with the metropolitan powers if its development potential is to be understood.

Second, dependency theory has tended to neglect the significance of the cyclical nature of capitalist accumulation on the world plane. Dependency theory was in itself a product of the 'long boom' from 1940 to 1970 — or, rather, of a particular phase of it. Between 1958 and 1965, a net flow of capital ran from the Third World to the USA of the order of US$16 billion, while there was also a growing net flow of funds out of the USA into Western Europe. From the standpoint of Latin America, where dependency theory was first developed, this seemed to portend a permanent exclusion from the industrialization process. This was theorized by A.G. Frank as merely an acceleration of the inveterate process of surplus transfer ('the contradiction of expropriation/appropriation'), and by Arghiri Emmanuel as the 'inevitable' movement of capital toward geographic areas of high demand for final products. It fell to Latin American writers such as F.H. Cardoso and O. Sunkel to point out, from the vantage point of Brazil in the late 1960s, that there had to be something wrong with this, because a veritable boom of mainly American private investment was occurring there. Cardoso, however, was too committed to the dependency framework to abandon it, coining the singularly unexplanatory term 'associated dependent development' to deal with the reality that Brazil and other rapidly industrializing Third World countries seemed to represent.[14] But the truth was that Brazil represented a phenomenon that would become more common as the European investment boom spent itself, and as Japanese competition in manufactures — coinciding with the first internationally synchronized recessions of 1970–71 and 1974–75 — created excess capacity, a surplus of investible capital and an attempt to restore profitability by shifting investment to areas that had higher rates of exploitation. In Africa, this tendency only affected significantly the Côte d'Ivoire and South Africa, but there seems to be no absolute reason why it should stop there.[15]

Third, dependency theory has not been much more alert to the active principles of imperialism than modernization theory had been. Although anti-imperialist in intention, its focus on the structures of economic subordination tended to direct attention away from the impact of great power military and geopolitical intervention. This may well have been more decisive, especially economic and military support for political clients and economic and military intervention against anti-imperialist forces, encouraging high levels of military spending and fostering military coups and wars. For instance, for the twenty-eight sub-Saharan African countries (excluding South Africa) for which estimates can be made, the real level of military spending rose fourfold between 1965 and 1977. In 1976 twenty sub-Saharan African countries were spending an average of 3.1% of GDP on their armies — not far below the level of the rich West European North Atlantic Treaty Organization (NATO) countries.[16] For countries that were mainly among the poorest in the world, these were crippling expenditures. An important part of this expenditure was directly due to imperialist military threats, as in post-1974 Mozambique and Angola, or was the outcome of complex great power manoeuvres, as in Ethiopia and Somalia. Mauritania was nearly crippled by its abortive colonial war in ex-Spanish Sahara, and Morocco too.[17] Fourteen of the thirty-six sub-Saharan African countries fought wars — civil or external — during the last twenty years; twenty (including some that had wars) experienced coups (some several).

All these considerations do not yield a more optimistic picture than that of the dependency theorists, but one that is more complex and less generalized. The insights of dependency theory must never be forgotten, but they must be separated from the utopian framework in which they have

[14] F.H. Cardoso, 'Dependent Capitalist Development in Latin America', *New Left Review* 74, July–August 1972, pp. 83–95.

[15] On the Côte d'Ivoire, see L.K. Mytelka, 'Direct Foreign Investment and Technological Choice in the Ivorian Textile and Wood Industries', in D. Ernst (ed.), 'Trends in International Transfer of Technology', special issue of *Vierteljahresberichte der Entwicklungslanderforschung*, March 1981.

[16] *SIPRI Yearbook* (New York: Oxford University Press, 1980), pp. 23, 31.

[17] By 1978, shortly before abandoning Tiris el-Gharbia to the Moroccans, Mauritania's military spending is said to have accounted for 60% of its overall budget (M. Bennoune, 'The Political Economy of Mauritania: Imperialism and Class Struggle', *Review of African Political Economy* 12, May–August 1978, p. 50). In 1978 its public debt was 138% of GNP. Morocco's was US$5.1 billion, up from US$711 million in 1970, and equivalent to 40% of GNP.

been cast. We must assess the African record in the context of the actual historical options; we must consider the actual historical forces — internal and external — at work in each country and region; we must take into account the cycle of accumulation and the reality of imperialist interventions of all kinds. The general result is likely to be much more complex, more diverse and much more dynamic than most scenarios in vogue today suggest.

This does not mean that no general patterns are to be found, although the variety that constitutes the African reality emerges from almost any classification one cares to make. In 1972 Samir Amin classified the African economies into three groups — colonial trade economies, plantation economies and labour reserve economies. For all its oversimplification, this schema had the great merit of grasping some essential distinctions between the different kinds of impact of colonialism on Africa, differences that imparted distinctively different dynamics to the development of the colonial economies, and it has been widely referred to since.[18] It is difficult to propose any equally useful simple typology for subsequent years. Table 5.2 classifies the African economies by two very rough measures of 'endowment', and by the

Table 5.2 Development strategy — a pattern of African development

	Sahel region		Mineral-exporting		Other	
Capitalist	Chad	(-1.4)	Niger	(-1.3)	Ghana	(-0.8)
	Senegal	(-0.2)	Zaïre	(0.7)	Uganda	(-0.2)
	Upper Volta	(0.3)	Liberia	(1.6)	Sierra Leone	(0.4)
	Sudan	(0.6)	Mauritania	(1.9)	Benin	(0.6)
	Mali	(1.1)	Togo	(3.6)	Central A.R.	(0.7)
			Nigeria	(3.7)	Rwanda	(1.5)
					Burundi	(2.1)
					Côte d'Ivoire	(2.4)
					Cameroon	(2.5)
					Kenya	(2.7)
					Malawi	(2.9)
					Lesotho	(6.0)
Socialist	Somalia	(-0.5)	Angola	(-2.1)	Mozambique	(0.1)
	Ethiopia	(1.3)	Guinea	(0.3)	Tanzania	(2.3)
			Zambia	(0.8)		
			Zimbabwe	(0.8)		
			Congo P.R.	(0.9)		

Source: World Bank, *Accelerated Development in Sub-Saharan Africa* (Washington, DC: World Bank, 1981). Figures in brackets refer to average annual rates of growth per capita 1960–79; the table excludes countries with fewer than 1 million population. For the Tanzanian figures, see note 35.

[18] S. Amin, 'Underdevelopment and Dependence in Black Africa: Origins and Contemporary Forms', *Journal of Modern African Studies* 10/4, 1972, pp. 503–24.

broad principles on which their development policies are based — the latter being important in themselves and also a rough indicator of the actual balance of social forces at work in the countries concerned.

African capitalism

If dependency is a new utopianism, a new 'philosophy of poverty' based on a preoccupation with the 'bad side' of periphery capitalism, does periphery capitalism have a 'good side' that has been neglected? This is the essential argument of one of the earliest and most consistent critics of dependency, Bill Warren. In his posthumous book, *Imperialism, Pioneer of Capitalism*, he developed the argument he had first advanced in 1973, that the post-war development of the Third World was actually a case of successful capitalist development — more rapid than in the industrialized countries, either historically or in the same post-war period; that the benefits of this capitalist growth were not restricted to the richest minority; that unemployment did not on the whole increase; that 'marginalization' is only a pejorative word for the process of increasing integration of new segments of the population into capitalist relations of production; that, in short, periphery capitalism performed its historic task of rapidly developing the productive capacity of the Third World.[19] What Warren did not stress, but which is equally central to his thesis, is that capitalist development is, and always has been, uneven, contradictory and costly in human terms. Its historically 'progressive' character springs from a 'grim, destructive, and oppressive logic'.[20] Warren did not argue that the capitalist development taking place in the Third World was 'nice' (although he considered that the middle-class anti-consumerist hostility to some of it was misguided and hypocritical).[21] He argued only that it was taking place; that no 'nicer' form of capitalist development (free from social dislocation, unevenness, exploitation, unequal exchange, inequality, oppression, wars and so on) has ever occurred or could occur; and that failure to acknowledge what was occurring could only handicap the masses of the Third World in their struggle to improve their lot, and

[19] Bill Warren, *Imperialism: Pioneer of Capitalism* (London: New Left Review Editions, 1980), chapters 7, 8; 'Imperialism and Capitalist Industrialisation', *New Left Review* 81, 1973, pp. 3–44.

[20] B. Beckman, 'Imperialism and Capitalist Transformation: Critique of a Kenyan Debate', *Review of African Political Economy* 19, 1980, pp. 46–62.

[21] See Warren, *Imperialism: Pioneer of Capitalism*, p. 249: 'The fact that purchase of consumer durables by low-income households occurs at the expense of public and perhaps other forms of consumption may be regarded as a distortion of resource allocation consequent to Western influence. But when did the poor ever know what was good for them? Would they actually opt for more collective (public) consumption if given the choice? No one knows, but it cannot be denied that most durable consumer goods — such as bicycles, sewing machines, motorbikes, radios and even television sets and refrigerators — significantly enhance the quality of life of poor households. It is only those who already possess such goods in abundance who feel it appropriate to suggest that it is undesirable for others to have them.'

eventually overthrow capitalism and replace it with communism. From this standpoint, dependency theory and the basic needs school appear as modern forms of 'critical-utopian' and 'bourgeois' socialism.[22]

There are various difficulties with Warren's position, but it has the great merit of forcing us to look again, as unsentimentally as possible, at the African record, which has caused so much distress to all but the hardest-nosed neo-classical economists working on Africa. If we omit the mineral-exporting economies, where sudden apparent increases in the growth rate may be due (although not necessarily) merely to changes in prices or the level of mining activity, we get the picture in Table 5.3.

It is first of all clear that the overall real growth of output was substantial

Table 5.3 Non-mineral-exporting capitalist countries in Africa

	Population 1979 (millions)	Average annual growth rate of GDP (%)	
		1960–70	1970–79
Sahel Region			
Chad	4.4	0.5	-0.2
Senegal	5.5	2.5	2.5
Sudan	17.9	1.3	4.3
Mali	6.8	3.3	5.0
Upper Volta	5.6	3.0	-0.1
Other			
Ghana	11.3	2.1	-0.1
Benin	3.4	2.6	3.3
Sierra Leone	3.4	4.3	1.6
Uganda	12.8	5.9	-0.4
Rwanda	4.9	2.7	4.1
Burundi	4.0	4.4	3.0
Kenya	15.3	6.0	6.5
Côte d'Ivoire	8.2	8.0	6.7
Cameroon	8.2	3.7	5.4
Malawi	5.8	4.9	6.3
Lesotho	1.3	4.6	7.0

Source: World Bank, *Accelerated Development in Sub-Saharan Africa* (Washington DC: World Bank, 1981); and World Bank, *World Development Report 1981* (Washington DC: World Bank, 1981).

[22] Of critical-utopian socialists, Marx and Engels wrote: 'The proletariat, as yet in its infancy, offers to them the spectacle of a class without any historical initiative... Only from the point of view of being the most suffering class does the proletariat exist for them... they are full of the most valuable materials for the enlightenment of the working class... [But their] proposals... are of a purely utopian character'; while the 'socialist bourgeois' were 'desirous of redressing grievances, in order to secure the continued existence of bourgeois society... They desire the existing state of society minus its revolutionary and disintegrating elements.' *The Communist Manifesto*, in Marx, *The Revolutions of 1848*, edited by D. Fernbach (Harmondsworth: Penguin, 1973), pp. 95, 93.

even in the 1970s, when world trade was faltering and the import costs of oil were rising so fast. It is true that some of the higher growth rates in the 1970s (Malawi's, for example) were due to favourable terms of trade. But this is irrelevant, since unfavourable terms of trade also account for some of the weaker growth rates (Malawi also suffered a 14% drop in its terms of trade between 1960 and 1970). It is also irrelevant that so much of the growth was in export agriculture, so long as agricultural export markets exist and domestic ones do not; and irrelevant that this expansion was achieved by planting more acreage rather than by higher yields on existing acres, so long as that is the most profitable way to expand output.[23] There was also a significant growth of production in general, and manufacturing in particular (with the exceptions of Sudan and Uganda), however 'inappropriate', dependent, capital-intensive and import-using it may have been. There were also substantial increases in life expectancy and in literacy, which would have been impossible if all the benefits of the growth that occurred had been entirely confined to the rich minority.

It is true that the levels of income achieved were very modest by world standards. But it is necessary for Western observers to calibrate their observations of Africa carefully. Although life expectancy at birth in sub-Saharan Africa in 1979 was only about forty-seven years, compared with seventy-four years in the industrialized countries, it was none the less eight years higher than in 1960, an increase of more than 20%. Literacy registered an even bigger advance, from about 5% to about 25% of the total adult population over the same period. Millions of Africans remained in abject poverty, but the poverty of many millions was less abject than it had been twenty years before. As Paul Kennedy noted in Ghana in 1976:

> What were luxury goods for one generation tend to become subsistence goods, or at least 'essential' commodities for the next. In Ghana today, the government is trying to control the distribution and therefore the price of a number of basic goods precisely because they are in such great demand. Some of these now seem to be regarded as 'essential' goods, for example, washing powder, toilet soap and tinned milk, whereas this was not the case 20 years ago. Other commodities, too, like transistor radios, bicycles, toothpaste and the possession of good clothes, in addition to every-day clothes, appear to be widely owned and used in the towns by the poorer sections of the community.[24]

It is also important to have an appropriate time frame. The inarticulate premiss of too much commentary is that Africa's industrial revolution should be capable of being accomplished in a generation or two. One study suggested, for example, that on the basis of recent trends it would take Kenya 201 years to begin to reduce the absolute number of people outside enu-

[23] Moreover, of the larger African countries, only the Côte d'Ivoire seems to have been very significantly helped by sustained above-average levels of official economic aid.

[24] P. Kennedy, 'Indigenous Capitalism in Ghana', *Review of African Political Economy* 8, January–April 1977, p. 29.

merated wage employment — with the implication that this showed the 'limited dynamism that marks recent Kenyan employment growth'.[25] But, apart from the fact that 'enumerated' employment is probably less than half of total wage employment in Kenya (a fact that itself indicates the spread of capitalist production relations ahead of the capacity of the state statistical service to record it), are two centuries really much too long for the conversion of the bulk of a country's labour force into wage earners (which, given the expected growth of population, is what this particular measure implies)? It also needs to be repeated that capitalist development is inherently uneven, both geographically and over time. For every success story in the African continent there are several failures. The chronic weakness of most of the Sahel economies, for instance, is as much due to the nature of capitalist development as are the success stories of countries such as the Côte d'Ivoire.[26]

What this implies is that African capitalism has accomplished more than dependency theory allows — although in a limited number of places, and at an immense cost in terms of wars, famines, corruption, oppression, cultural deprivation, militarism and foreign domination. But even this claim may be questioned. Are the advances that capitalism appears to have made in some places capable of being sustained? Are the best records of the 1960s and 1970s a reflection of easy options — more extensive agriculture, import-reproducing industrialization, and so on — that are now largely exhausted and that cannot be replaced? Are the gains of these years — such as they are — essentially precarious and about to be swallowed up in the inexorable growth of the African populations?

The debate on this question has so far revolved largely around the two fastest growing African economies (other than those with strong mineral exports), namely, Côte d'Ivoire and Kenya, the assumption being that if their growth cannot be sustained, the prospects for relatively poorly endowed African economies to develop further within capitalist relations of production must seem slight.[27] Both countries embarked at independence on an explicitly capitalist development policy, although not described as such; Kenya called its strategy African socialism, while Côte d'Ivoire called

[25] S. Langdon, citing F. Stewart in 'Industry and Capitalism in Kenya: Contributions to a Debate', paper prepared for the Conference on the African Bourgeoisie, Dakar, 1980, mimeo, pp. 16–17.

[26] Capitalism does not cause the rains to fail — as far as we know — but it causes changes in land use that expose people to the risk of disaster when the rains fail, and it finds no profit in most developments that might permit the Sahel populations to re-establish their traditional symbiosis with their environment. See C. Meillassoux, 'Development of Exploitation: Is the Sahel Famine Good Business?' *Review of African Political Economy* 1, 1974, pp. 27–33; A. Baird *et al.*, *Towards an Explanation of Disaster Proneness*, University of Bradford Disaster Research Unit, Occasional Paper no. 11, August 1975.

[27] As testimony to what has been said earlier about the dynamic nature of African development, the Côte d'Ivoire was expected to become self-sufficient in oil by 1982, and seemed likely to become a mineral exporter thereafter.

it 'planned, contractual and controlled liberalism'.[28] Both opened their doors to foreign investment and encouraged domestic rural and commercial entrepreneurship. Kenya had the advantage of a relatively large settler enclave and an Asian-owned commercial sector, both of which could be Africanized. Côte d'Ivoire had received a much more modest stimulus from settler agriculture, but adopted a peculiarly close relationship with France, including very high levels of technical and capital assistance; it was the pivot of France's continuing commercial dominion of the former French West Africa.[29]

Those who see the Kenyan and Ivorian experience as evidence that capitalism can develop Africa, point to the relatively high rates of growth that have been achieved in these countries on the basis of an alliance between a local capitalist class and foreign capital. Dependency theorists, on the other hand, consider this growth precarious and as having been achieved partly by extending agricultural production in ways that have reached their limits and partly by making most of the population pay more for their necessities through the protection given to a highly inefficient and often socially undesirable import-reproducing manufacturing industry. They see industrial employment rising more slowly than industrial output, and foreign exchange constraints eventually halting the advance that was achieved in the 1960s and 1970s.

In spite of its interest and importance, this debate has struck many observers as inconclusive. The record of growth does have a lot to do with favourable commodity price movements, the expansion of world markets in the 1960s, and French policy favouring Ivorian industrialization within the ambit of France's own domestic plan. On the other hand, it also has to do with increased commercialization of farming, rapidly expanding informal sectors in manufacturing, large transfers of know-how and so on.

The real issue, however, is whether the capitalist expansion of these economies that undoubtedly occurred is necessarily fated to come to a halt, not whether it is fated to encounter new problems, as it surely is. The dependency school takes these problems and elevates them to the status of a theory of 'blockage'. Their opponents see the problems as inherent in capitalist development, but see no reason to suppose that they cannot be overcome. For instance, the fact that imported branch-plant technology tends to minimize the share of manufacturing inputs obtained locally is frequently cited as a major reason for industrialization leading to foreign exchange constraints and not to the expansion of linked domestic industries. This is true, but it is not impossible for African host governments to force such plants

[28] B. Campbell, 'Capital Accumulation and the Post-Colonial State in the Ivory Coast', paper presented to the panel on 'Bureaucratic Bourgeoisie or Ruling Elite', African Studies Association Conference, Philadelphia, 1980, p. 10.

[29] In 1978 gross official aid inflows to the Côte d'Ivoire were US$122 per capita, compared with US$16 for Kenya, and an average for sub-Saharan Africa of US$31.

to increase progressively over time the proportion of locally produced inputs that they buy, as has been done in Latin America. Another oft-cited problem is the falling rates of growth of agricultural output, leading (as in Kenya and Nigeria) to rising food imports. This constraint too can be tackled by agricultural reorganization to raise farm productivity. Such possibilities, however, depend greatly on the exercise of state power, and it is here that the question of the class structures of these countries becomes especially important, since the exercise of state power expresses the balance of class forces in the dominant alliance or bloc, and the counterweight of opposing classes.

In this connection, the so-called national bourgeoisie in Kenya and the so-called 'planter' class in Côte d'Ivoire have been the chief focus of attention. Up to a point, this focus is correct. The nature of the 'class of capital' is bound to be important for the nature and rate of capitalist development. Its significance, however, is easily misunderstood. Dependency theorists tend to look at the matter in the following way. Foreign capitalists have no interest in developing the periphery, but only in exploiting it for the profit of their companies at the centre. Domestic or peripheral capitalists have an interest in developing the periphery (i.e. maximizing investment in their own countries), but lack the means and the know-how. Sometimes it is said that they lack the will as well. This perception of the African businessman as essentially a comprador, content to live parasitically as a commission agent for foreign capital, was crystallized in Frantz Fanon's savage appraisal:

> This native bourgeoisie, which has adopted unreservedly and with enthusiasm the ways of thinking characteristic of the mother country...will realize, with its mouth watering, that it lacks something essential to a bourgeoisie: money...If the government gives it enough time and opportunity, this bourgeoisie will manage to put away enough money to stiffen its domination. But it will always reveal itself as incapable of giving birth to an authentic bourgeois society with all the economic and industrial consequences which this entails.[30]

Fanon was referring to something real. But is it generic, as he implied? Is it equally true of all African countries? Is it permanently true anywhere? Is it as true today as it was in the Africa Fanon knew at the end of the 1950s? There can be little doubt that in some African countries bourgeoisies have begun to form that have at least some capacity to organize domestic capitalist production. Their emergence, however, was strongly conditioned by the histories of the particular ex-colonies in which they have appeared. It is an oversimplification, but not a very misleading one, to say that in Kenya and Côte d'Ivoire their advance was assisted in different ways by white settler interests, which pioneered a degree of internally orientated development. This pattern of development both stimulated the emergence of more advanced African producers and at the same time discriminated against them, finally driving them to take a leading role in their respective nationalist

[30] Frantz Fanon, *The Wretched of the Earth* (Harmondsworth: Penguin, 1967), pp. 143–44.

movements. Elsewhere in West Africa, the colonial trade economies produced, from an early stage, indigenous classes of merchant capitalists and petty traders. These different trajectories have left their marks on the different African capitalist classes, especially in the degree to which they are orientated to production rather than trade. These differences, however, can be exaggerated; the 'logic of capital' no doubt works progressively to reduce them.[31] There seems no good reason to suppose that Nigeria will for ever lack for manufacturing entrepreneurs, and Paul Kennedy claims not to have found them scarce in Ghana in the late 1960s. It is equally a mistake to think, in any case, that foreign capital is not interested in developing the periphery. What capital is interested in is not development, but its own expansion. When the conditions exist for profitable productive investment in the periphery, both foreign and domestic capital (no longer salivating for lack of money) become interested in it.

The emergence of such conditions has something to do with stages of growth — the commercialization of agriculture, the separation of workers from the land, the expansion of internal demand. But it also has to do with politics, and it is the political power of domestic capital that is at least as important as its capacity to organize production on its own. When circumstances permit an indigenous capitalist class to establish itself effectively in power, as in Côte d'Ivoire and Kenya, the conditions for capitalist development at the hands of both foreign and domestic capital are enormously enhanced. Such power is, of course, never limitless; in Côte d'Ivoire, for example, the mutual dependence of the Ivorian plantocracy, French capital and the French state was reflected in the extremely cautious — but profitable — strategy pursued by Houphouët-Boigny in advancing national interests *vis-à-vis* French interests. But the emergence of African states capable of sustaining the conditions for capitalist growth — a measure of reliability in public administration, social control without repression so severe as to jeopardize stability, a non-arbitrary application of commercial and criminal law, and so on — is importantly conditioned by the political power exercised by a domestic class with substantial capital of its own at stake.

The real measure of the importance of a politically cohesive production-orientated domestic class of capital is the fate of various countries that for various reasons have lacked one hitherto: Zaïre, for example, and Nigeria, both rescued to a lesser or a greater degree by mineral exports; Mauritania and Niger, which seem likely to forfeit much of the benefit that their mineral exports might bring at the hands of an effective local bourgeoisie; above all, Uganda. Here, the African capitalist class was politically divided, and the Baganda, who made up the most important section of it, were fatally

[31] None the less, the embodiment of a class's economic and political capacities in its familial and social institutions and culture does exercise an independent influence. Apart from relatively early studies such as those by Samir Amin, P. Marris and A. Somerset, and S. Kaplow, this topic awaits proper investigation in Africa.

ambivalent — economically, in their commitment to accumulation (as a result of the semi-feudal relations of production in Buganda), and politically, in their inability to transcend the parochial confines of Ganda traditionalism, as was needed in order for them to secure a political basis for capital accumulation.

The Ugandan experience under Amin from 1971 to 1979 provides, in fact, a sort of negative object lesson in the substantiality of capitalist development in Africa. In office was a dictator whose political power rested at first primarily, and later exclusively, on force. Revenues were channelled increasingly into military expenditures, the trading sector was in effect seized and parcelled out among army officers, soldiers and their associates; the civil state apparatus, including its economic management branches, was decimated and demoralized. The conditions for simple reproduction, let alone expanded reproduction, were destroyed. The whole economic structure unravelled. Primary crop production continued, but on a reduced scale, and less and less produce was traded officially. By 1979 prices had risen about 2,000% above 1971 levels. Moonlighting and black market trading became essential for salaried workers if they were to stay alive. The advance earlier achieved by capitalism in Uganda — which had an average annual increase of GDP of 5.9%, or 2.2% per capita, in the 1960s — can be judged from the difference between 1970 and 1978–79 in these figures, as shown in Table 5.4. At the same time, it was the particular unevenness of capitalist development in Uganda that contributed decisively to the catastrophe of the Amin regime.[32]

Two particular kinds of constraint on future growth loom large in the

Table 5.4 Output of selected products in Uganda 1978–79, as a percentage of 1970

Fabrics	55
Soap	9
Matches	16
Cement	38
Corrugated iron sheets	7
Sugar	8
Cotton (officially marketed)	14
Coffee (officially marketed)	45
Blister copper	14
Goods forwarded by rail	13

Source: Commonwealth Secretariat, *The Rehabilitation of the Economy of Uganda* (London: Commonwealth Secretariat, 1979).

[32] See E.A. Brett, 'The Political Economy of General Amin', *IDS Bulletin* 1/7, April 1975, pp. 15–22.

thinking of most dependency writers on Africa (including the basic needs school): the size of the internal market and agricultural productivity. These constraints are real. The limited size of most African domestic markets — even Nigeria's, after the oil boom — sets severe limits to successful manufacturing development, especially using imported technology geared to large production runs. Food production per capita was also falling in African economies generally in the late 1970s, posing a serious threat to further non-agricultural advance. Côte d'Ivoire suffered from the first of these constraints, Kenya from both. The former tackled the problem of market size by trying to become an exporter (of wood veneers and textiles) to Europe under the terms of the Lomé Convention. Kenya could not easily follow this route, because its textile firms were not European-owned but largely Japanese and Indian, whereas it suited the French firms in Côte d'Ivoire to restructure their production there to serve their European markets, as the latter grew more competitive. Does this mean that Kenya's industrial growth will be blocked? To suppose so is to suppose that neither the multinational corporations nor the Kenyan business class associated with them is capable of devising new strategies to deal with the problem of market size — from new tax regimes to new foreign policy initiatives to restore the East African regional market, new initiatives in Europe and so on. It is not obvious that this supposition is well founded.

Similarly in relation to agriculture. The problem of food production is technically soluble; according to John Carlsen, writing in 1980:

> in relative terms, very little irrigation is being used, very few chemical inputs are used and almost no mechanisation has taken place in Kenyan agriculture...the easy options of Kenyan agricultural development have still not been fully utilized. High-yielding varieties of maize, millet and sorghum have still only been adopted by a minority of the peasants...Also, the potential of high-yielding dairy cattle has not yet been fully exploited...The potential of mixed farming has hardly been realized in Kenya outside Central Province.[33]

What is ultimately in question is the political possibility of reorganizing the relations of production in agriculture so as to raise productivity — implying, perhaps, an accelerated process of land concentration in the hands of a few capitalist farmers, and an accelerated increase of landlessness, or the development of state-subsidized irrigation for large-scale agriculture, and so on. There is no question that these are difficult hurdles to jump, but some African countries will find ways of doing so. On the other hand, only a limited number will, at least in the foreseeable future. It would be as mistaken to think that capitalism is in the process of developing all the countries of Africa as it is to suppose that it has not developed, and cannot develop, any of it.

[33] John Carlsen, *Economic and Social Transformation in Rural Kenya* (Uppsala: Scandinavian Institute of African Studies , 1980), pp. 223–24.

'Actually existing' African socialism

Long before the academic debate about capitalism and dependency had even started in Africa, many African countries had come to doubt either their ability to develop within capitalist production relations, or the desirability of doing so, or both. A strong socialist strand existed in the nationalist movements of many, if not most, African countries, and the post-independence experience of Africa provided it with plenty of reinforcement.

We may distinguish at least four different groups of African socialist regimes: (i) the regimes more or less directly descended from the Rassemblement Démocratique Africain (RDA), linked to the French Communist and Socialist parties for most of the 1940s — Guinea, Mali (1960–68), and the Congo People's Republic; (ii) regimes that chose a 'socialist' option out of a conviction, based on post-independence experience, that capitalist development was impossible, or not in the popular interest — Ghana (1960–66), Tanzania, Zambia, Somalia and Benin;[34] (iii) Ethiopia (after 1974), whose 'socialist' option was a complex blend of the antifeudal movement, especially among the intelligentsia, and the Russian/Cuban military alliance; and (iv) regimes formed in anticolonial struggles — Guinea-Bissau, Mozambique, Angola and Zimbabwe.

Obviously, there was as much variety in African socialism as it actually existed (not as it hoped to become) as there was in African capitalism: from enormously differing endowments to widely differing conceptions of the transition to socialism; from the rigid centralism of Touré's Guinea (at least until 1979) to the comparatively open and certainly undoctrinaire regime of Nyerere's Tanzania; from Ethiopia's enforced dependence on the USSR to Zimbabwe's enforced dependence on the West. There remained, however, clear areas of convergence: public ownership of strategic industries; state or co-operative trading; limits on foreign investment; trade and aid orientations that included Comecon economies or China, or both; restraints on the emergence of indigenous capitalism; efforts to equalize private consumption; efforts to promote collective or co-operative farming; and, most important of all in the long run, efforts to foster popular power in local as well as national institutions.

The economic growth record of the socialist countries is particularly hard to assess, in view of the additional obstacles placed in their way by imperialism — from France's attempt to sabotage the Guinean economy after 1958 to the subsequent abortive invasion of Conakry in 1970; from

[34] We exclude Sudan, and should perhaps exclude others, whose socialism comes too close to the purely rhetorical variety of Tolbert's Liberia or the last days of Bokassa in the Central African Empire: 'The Central African Emperor has decided to adopt the revolutionary and authentically African and nationalist ideas of the Libyan Arab Jamahirya.' R. Delpey, *La Manipulation*, cited by M. Boli Richard in the *Le Monde* section of the *Manchester Guardian Weekly*, 5 April 1981.

the aid penalties imposed on Ghana and Tanzania after their exercise of a 'socialist' option to the assassination of outstanding leaders such as Mondlane in 1969 and Cabral in 1973; and the burden imposed on Mozambique and Angola by South African military attacks. In the case of Zimbabwe, the British-dictated constitutional limitations on nationalization, the political power of the white bureaucracy and army units, the economic weight of the white working class and foreign capital and the South African threat all combined to make it very hard to assess the practical potential of the regime's 'socialism'.

But, when all allowances are made, the general economic record of the African socialist countries also disappointed sympathetic observers. Several of them succumbed to antisocialist coups (Mali, Ghana and, eventually, Guinea-Bissau). In others, socialism reduced itself largely to the nationalization of major foreign assets, combined with redistributive, welfarist and normally agrarian rhetoric, more or less imperfectly matched by performance on the part of the single party or military leadership (e.g. Zambian 'humanism'). Others never really emerged from the trauma of the liberation struggle. In the socialist camp, Tanzania stood out, like the Côte d'Ivoire and Kenya in the capitalist camp, as the focus of debate, because of its persisting socialist initiative and a relatively well-sustained growth rate unfavoured by natural resource endowment.[35] What light does the Tanzanian debate shed on the possibilities of the socialist option?

As is well known, the Tanzanian initiatives took five principal forms: (i) democratization; (ii) nationalization; (iii) income equalization and measures to prevent the leadership from turning itself into a bourgeois class; (iv) collective farming; and (v) national economic 'self-reliance'. By the standards of many African countries, Tanzanian efforts at democratic one-party government were genuine, and the tendency to arbitrary administration was contained. But only the regime's most uncritical apologists would maintain that elections offered the voters any serious influence on policy development or contributed significantly to raising their level of political consciousness, while attempts to introduce worker self-management, in 1971 through plant-level workers' councils, and from 1976 through village councils, were largely neutralized by the growing bureaucratism and authoritarianism of the state apparatus. Nationalization was also a mixed success; on balance, it seems probable that the state-run industrial sector was relatively efficient and the state-run commercial sector rather inefficient (leading to some instances of reprivatization). A 'leadership code' successfully blocked senior party and state officials from engaging in capitalist

[35] There is an apparent discrepancy between the decline in the volume of agricultural production that occurred in Tanzania in the mid-1970s and the growth rates of GDP shown in the national accounts and recorded in Table 5.2 above (see *Accelerated Development in Sub-Saharan Africa*, (Washington DC: World Bank, 1981), p. 187). There may also be other reasons for the high overall growth rates apparently achieved, but it seems more likely that the figures are too high.

accumulation, income differentials between state employees were strikingly reduced, and the scope for private capital accumulation in commerce and farming was severely limited by the nationalization measures, credit policy and so on. Socialist farming, however — evidently crucial, politically and socially as well as economically, in a country where 83% of the workforce is in agriculture — was unsuccessful as a voluntary programme, and in 1973 was largely abandoned, to be replaced by forced 'villagization', under — in theory, at least — increasingly close state supervision. Combined with severe droughts, this led in 1974 to a catastrophic fall in agricultural output, including food output, from which the economy has not yet fully recovered. Compounded by the oil price increases in 1973, and then by the Ugandan war of 1978–79, this drop in output led to a period of intense austerity, increased reliance on authority to secure agricultural production, and the abandonment of at least a significant part of the hoped-for financial and commercial national self-reliance. As a result, foreign borrowing and technical and policy inputs from the World Bank and Western aid donors increased dramatically.[36]

Two broad schools of thought can be discerned in the debate about this experience. One holds that a society is 'in transition to socialism if...[it is] ...deliberately reshaping itself along more egalitarian and more participatory lines and...promoting co-operative and non-acquisitive motivations among men and women in their economic and other inter-relationships'.[37] In this view, Tanzania grappled admirably with the irreducible difficulties that stand in the way of all such endeavours. The other school holds that this conception of socialism is idealist — that is, it sees socialism as a matter of values and intentions (in Nyerere's own words, 'socialism is an attitude of mind') divorced from the realities of the production relations that actually exist. These relations, inherited from colonialism, are capitalist, and the second school considers that capitalism has a logic that cannot be indefinitely defied without a fundamental crisis of production. It considers also that the crisis of the mid-1970s, while aggravated by the drought, was ultimately attributable to a failure to confront the basic contradiction between two principles on which productivity can be increased — either the drive for individual accumulation in a competitive market *or* the drive for accumulation by self-determining co-operative groups within a democratic co-operative commonwealth.

The initial form taken by this criticism was that the regime's socialism would prove empty unless it involved the mobilization of the poorest peasants and workers against the rich peasants and their employers, including the state bureaucracy (identified as a bureaucratic bourgeoisie). Subsequently, S. Mueller, in a forceful critique inspired by the post-dependency

[36] Tanzania's external public debt rose from US$248 million in 1970 to US$1,095 million in 1978. *World Development Report* 1980, p. 138.

[37] *Towards Socialism in Tanzania*, edited by B.U. Mwansasu and R.C. Pratt (Toronto: University of Toronto Press, 1979), p. 6.

debate, emphasized the close parallel between Tanzanian policy and that of the populists (Narodniks) against whom Lenin polemicized at the turn of the century in Russia. The Tanzanian syndrome, she argued, has a distinctive logic: in the name of socialism, the development of an entrepreneurial class, whether in industry or agriculture, is thwarted; but peasant production, the basis of all economic advance, is thereby confined to the household plot and the hoe. No division of labour is possible; no significant productivity increase can be assured by state-provided inputs of seeds or fertilizers, given the heavy disincentives to individual advancement (state control of land and credit, opposition to the expansion of full-time wage labour and so on). The result is stagnant or declining output, provoking fresh efforts of state control (appointment of village managers and village management technicians, reintroduction of minimum acreage laws from the colonial period, restrictions on population movement, corvée labour and so on) and leading ultimately to efforts to break the bottle-neck by the polar opposite of 'African socialism' — large-scale capitalist farming financed by international capital (the World Bank, in particular).

> Narodism stifles class formation and gives birth to a repressive state. It cannot be the vehicle for a transition to socialism, and no such process of transition is occurring in Tanzania. Narodism in that country has instead been the vehicle for pauperisation both at the economic and political levels...Here, socialism has not triumphed over capitalism...In Tanzania one is only viewing capitalism in its 'least developed' and 'worst forms'...With the extraction of relative surplus value precluded under this system, compulsion is the only way the state can squeeze greater surpluses from its poor and demoralised peasantry. This necessitates an increasingly repressive political and administrative system, which in turn reduces the opportunities for popular expression and political participation.[38]

It is difficult not to accept the main thrust of this analysis, in spite of its polemical tone and rather one-sided character — difficult not to recognize, through it, the theoretical basis for an understanding of 'actually existing' African socialism,[39] of culturally and practically diverse experiences, from Tanzania to Guinea and Guinea-Bissau. If the avatar of African capitalism is a corrupt, arbitrary and despotic regime presiding over economic regression, the avatar of African socialism is an increasingly hollow

[38] S. Mueller, 'Retarded Capitalism in Tanzania', *Socialist Register 1980*, pp. 220–21. Comparisons of any sort between contemporary Africa and late Czarist Russia are apt to seem far-fetched at first sight, but the similarities between the thought of the Russian populists on the fate of the peasantry after the emancipation of the serfs and that of some contemporary theorists on the African peasantry are striking (for a recent example, see G. Hyden, *Beyond Ujamaa* (Berkeley: University of California Press, 1980)). Lenin's study, *The Development of Capitalism in Russia*, provides the starting-point for the critique advanced by Mueller, as it does for the work of M. Cowen and others on the very different political development in Kenya.

[39] The phrase is Rudolf Bahro's, in *The Alternative in Eastern Europe* (London: New Left Books, 1978).

and parasitic form of domination by the state *apparat*, again leading to economic regression. Just as 'in the colonies, there are more bourgeois-minded people than bourgeois', in the African socialist regimes, there tend to be more socialists than there is socialism. And not all that many socialists either, in some cases.[40]

But it is a mistake to dismiss the Tanzanian experience, or any other socialist project in Africa, as no more than a mutation of backward capitalism. Any such judgement involves an element of determinism that echoes that of the dependency school and must be rejected. To idealize actually existing socialism is useless; to minimize its significance is an equal and opposite error to that of failing to recognize the development that capitalism, in a few places, has brought about.

The accomplishments of African socialism are not, on the whole, to be measured in terms of growth rates. They are primarily social and political, above all in having posed the question of the form within which development is to occur, in having made it comprehensible to ordinary people that they do have collective historical choices which they may try to exercise if they will. The achievement of the Tanzanians and the Ghanaians in this respect is epochal. One can also say that the accomplishments of actually existing African socialism lie partly in its failures; to paraphrase Marx, what succumbed in these failures was not African socialism but the 'persons, illusions, conceptions, projects, from which the idea of socialism in Africa was not free, from which it could be freed...only by a series of defeats'.[41]

As with African capitalism, an adequate perspective on African socialism must be long-term and continental in scope. The Nkrumahist illusions of 'state socialist' industrialization, which succumbed to the 1966 coup in Ghana, and the populist illusions of co-operative farming without class conflict, which succumbed to the forced villagization decision of 1974 in Tanzania, are lessons of history that belong to all the African peoples. The immediate beneficiaries of such failures are often implausible collections of military officers, posing as councils of national redemption; but the long-run legatees are those in all African countries who have contemplated these experiences and are increasingly determined that the people as a whole should profit from them. To doubt this is to ignore the growth that has occurred, in spite of all the setbacks, in the awareness and understanding of labour leaders, students and farmers in Africa over the last two decades, and to underrate the will and capacity for historical action that has been shown by ordinary people in innumerable boycotts, 'hold-ups', strikes, 'riots', and rebellions in all regions of the continent, to an extent that has not yet been adequately documented.

[40] The saying is quoted in Roger Genoud's important early study, *Nationalism and Economic Development in Ghana* (New York: Praeger, 1969), p. 52.

[41] See Marx, 'The Class Struggles in France', in *Surveys from Exile*, edited by D. Fernbach (Harmondsworth: Penguin, 1973), p. 35.

Political perspectives

Implicit in these remarks are analytical and political judgements that should be candidly stated. Analytically, they rest on the assumption, which is still occasionally denied, that classes are forming in Africa, whose struggles will shape the future development of the continent. Politically, what has been said here rests on the view that the underclasses of Africa, whose logical tendency is toward socialism, must be supported in their struggles.

Neither of these judgements can be adequately defended, or perhaps even explained, in a brief compass. The formation of classes is bound up with the mere fact of capitalist penetration throughout the continent. To the extent that this process is incomplete, embryonic and uneven, the development of the African classes is too. This does not make them less real. Those who doubt the historical significance of classes in Africa tend to contrast them with the classes of Europe, attributing to the latter a homogeneity and cohesion they have never possessed, and which they lacked most spectacularly when Marx first defined the class character of capitalism in the late 1840s. It is true that the African classes differ in many ways from classes in contemporary Europe or North America. The much earlier stage of capitalist development accounts for part of the difference. The specific histories of these countries, the forms taken by the prior stages of class struggle and the distinctive class cultures and ideologies resulting from them are equally important.

For example, the central and still very recent role played by the colonial state and foreign capital affected the genesis of the African capitalist, and especially the African industrial capitalist, to the extent that many observers have doubted his very existence — until one day he is encountered, buying out an import agency in Accra, establishing a factory in Nairobi or owning a block of apartments in London. At the other end of the scale, migrant African workers, such as the cocoa and coffee farm-workers in Ghana, Côte d'Ivoire and Buganda, have been 'proletarianized' in circumstances that tend to neutralize them politically, like Turkish workers in Germany or Mexican workers in California. But the increasingly stabilized workforces of the factories in Zaria or Nairobi or the gold mines in Ghana have accumulated traditions of political action that correspond quite closely to the 'corporate' level of consciousness that marked the earliest stage of the formation of the working classes in Europe and elsewhere. The nature of post-colonial electoral mobilization by parochial parties, now largely discredited, has left its mark in a fragmented political culture, in which class ties compete with multiple other forms of consciousness — ethnic, local, religious and so on. Yet the economic evolution of the African countries has tended to throw their class character into sharper and sharper relief.

The large literature on the development of classes in Africa has not so far yielded much agreement on the specific character and tendency of class formation in Africa generally. The verdict is still open on whether the higher state bureaucracy constitutes in itself a new kind of dominant class; whether state employees as a whole form a distinct, privileged class or constitute 'aristocratic' strata within other classes; under what conditions peasants, who still form the great majority of the African populations, have acquired or may acquire a radical consciousness and the capacity for sustained political mobilization; and what forces determine the evolution of the balance between class consciousness and ethnic, regional or religious consciousness.[42] But few would now deny that class forces are critical for the fate of development in every African country or that every development strategy has an inescapable class character.

As for the political standpoint from which African development has been considered here, it is, in intention at least, one of solidarity with the African masses who are impoverished and oppressed and who are in various ways struggling to overcome this — a protracted, partly conscious, partly unconscious collective historical aspiration and movement best captured in the expression 'the African revolution'. Perhaps the most valid of all the insights of the dependency school is the recognition of the immense costs that people in Africa (and other regions of the Third World) have been paying for the capitalist development of, at best, some parts of the continent, in the interests of a narrow alliance of domestic and foreign capitalists. This process has generated many forms of resistance and, in the long run, this resistance is the basis on which any genuinely popular alternative development strategy will have to be founded. The weaknesses of the first African 'socialisms' are, from this point of view, not really so different from those of the early European socialisms. They are, or were, the product of the masses' first encounters with capitalism, and perhaps were unavoidably conceived and implemented primarily by the educated middle class, whose aspirations on behalf of the masses tended to be counterbalanced by their willingness (often perceived as an inescapable necessity) to decide everything on their behalf. These utopian and often gravely compromised socialisms, I have argued, are not to be dismissed, but they cannot command unqualified support either. A commitment to the emancipation of Africa involves judgements as to he forces and projects that best represent the interests of the masses, that are most democratic, judgements that are as difficult to make in Africa as they are everywhere else. But the mistakes of those who really try to answer these questions will more often than not be generously forgiven.

The Western observer who adopts this position is often challenged to say what he or she would recommend 'instead' of what this or that regime, capitalist or socialist, is doing or has done. This sounds like a reasonable

[42] A leading proponent of the view that the higher bureaucracy of the African state is forming itself into a dominant class has been J.S. Saul (see, for example, his *State and Revolution in Eastern Africa* (New York: Monthly Review Press, 1980)). For convenient references to the now enormous literature on class in Africa, see *Review of African Political Economy* 19 (September-December 1980), special issue on Consciousness and Class.

demand, designed as it is to curb the pretensions of armchair criticism. Yet what the African revolution needs is not so much recommendations as support. Outsiders can contribute usefully to African policy-making, but too many of these efforts are vitiated by the weakness of the popular forces on which most progressive policies ultimately depend for their success — and this weakness is aggravated by the failure of too many in the West to extend to the African revolution even the most elementary forms of support. How many of these challengers have worked for the release of trade unionists detained in various African countries or protested at the shooting of students in Zaïre or the hanging of Zimbabwean guerrillas by the illegal Smith regime — or even against the lending of their own money by their own banks to the repressive racist state of South Africa?

This discussion may seem to have wandered far from the problem of economic development, but such is not really the case. The African peoples are extremely poor. Even if external conditions are favourable, most African regimes will be unable to do much more than moderate the poverty of most of their populations in the next generation. In reality, it is more likely that their fragile economies will be subjected to new stresses, arising from the global contradictions of capitalism and imperialism. Given that the productive capacity exists in the Western world to relieve this poverty entirely and to initiate rapid increases in African productivity, this human waste is abominable as well as dangerous.[43] Yet in the last analysis, a solution will not come from anywhere but Africa. Those who envisage only long-term stagnation and decline forget that it is out of the crises and struggles to which these inevitably lead that new social forces, capable of new solutions, are radually emerging.

[43] This is the main burden of the Brandt Commission Report, *North–South: A Programme for Survival*; see pp. 12–17 and *passim*.

The State & the Crisis of Simple Commodity Production in Africa

Sub-Saharan Africa has over 400 million people, many of whose per capita incomes, already among the world's lowest, have stopped growing and in many areas may well fall. Its economies are overwhelmingly agricultural, and yet, according to the World Bank, agricultural output is rising more slowly than population growth, which is currently growing at an average of 3.3% per annum (leading to a projected population of a billion by the year 2010). The failure of output to keep pace with population growth has compounded the difficulties posed for Africa by the world economic crisis and the monetarist response to it. For sub-Saharan Africa as a whole, the World Bank predicts that a third of all food requirements will have to be imported by the year 2000, while foreign debt service already absorbs about 15% of all export earnings.

Accompanying and aggravating the economic crisis is a general social and political crisis. Military coups have occurred in a majority of countries in sub-Saharan Africa, and more than half have military or quasi-military governments. More state spending is devoted, on average, to the military than to either agricultural development or health education. Basic personal security is often lacking, and in some countries corruption and extortion by state officials is commonplace. Where this is not the case, inefficiency is often damagingly severe, so that fertilizer arrives too late to be used, trucks arrive too late to remove the harvest, salaries are overdue for months so that administrators cease to function and so on. The most popular interpretation of this state of affairs is that it is due to misguided development policies. The new orthodoxy in the so-called 'development community' proposes that the African states fell into the wrong hands at independence, which has brought about an inefficient use of resources; the remedy is to bring about a contraction of the state and allow market forces to operate. This line of thought (although with a less market-orientated conclusion) is exemplified in Richard Sandbrook's book, *The Politics of Africa's Economic*

*First published in *IDS Bulletin* 18/3, 1987.

Stagnation.[1] Sandbrook argues that the problem is that the African states have failed to perform the task that capitalist growth has always required from the state, because they have reflected a precapitalist social structure. They have been predominantly 'neo-patrimonial' in character, based on the personal domination of an individual leader, who uses the country's resources primarily as loot for rewarding his loyal followers. Less extreme versions of this thesis see the problem in terms of 'urban bias' on the part of state bureaucrats and of politicians sensitive to the demands of primarily urban political elements.[2]

Against this prevailing orthodoxy, Bienefeld has persuasively argued that it grossly understates the responsibility of the same 'development community' for pursuing policies that over the thirty years since independence have rendered Africa acutely vulnerable to pauperization by the nature of its place in the world economy, and that increasing the scope of operation of market forces will only make the crisis more acute and intractable.[3] According to this view, the problem is that, under constant pressure from multilateral and bilateral aid donors, the African states have failed to pursue 'coherent' strategies to create a degree of insulation of their economies from world market forces. The solution is seen as lying in more coherent development strategies based on the long-term national interests of the African peoples.

While there is more to be said for the latter view than the former, the most obvious general conclusion to be drawn from this debate is that it is unwise to generalize. There are African states that are little more than (badly) organized gangs for looting the national surplus, and there are others that have accomplished much more than the antistate arguments of both the right and some of the left in the debate recognize.[4] None the less, regarding the causes of the African development crisis there is a widely distributed feature of African production which deserves attention, namely the social relations of agricultural production, which remain predominantly those of individual households using family labour on small plots which they hold in some form of ownership.

For a hundred years or more, this system of 'simple commodity production' expanded output rather steadily and substantially. In the twenty years from 1960 to 1979 alone, for example, the volume of merchandise exports from sub-Saharan Africa more than doubled, and the bulk of this came from peasant households. The rise in living standards in Africa, reflected in

[1] Richard Sandbrook, *The Politics of Africa's Economic Stagnation* (Cambridge: Cambridge University Press, 1985).

[2] See, for example, Robert Bates, *Markets and States in Tropical Africa* (Berkeley: University of California Press, 1981).

[3] M.A. Bienefeld, 'The Political Economy of Sub-Saharan Africa's Crisis', *Review of African Political Economy* 43, 1988, pp. 68–87.

[4] For example, Lord Bauer, *Equality, the Third World and Economic Delusion* (London: Methuen, 1981); and, on the the left, E.A. Brett, 'State Power and Economic Inefficiency: Explaining Political Failure in Africa', *IDS Bulletin* 17/1, 1986, pp. 22–29.

major advances in education and life expectancy, as well as in personal incomes, was built on the same economic foundation. Why, then, should we single out the system of simple commodity production as a general cause of the crisis? The reason has to do with the special tasks that the state is called on to perform if agricultural productivity is to be raised under this system of production.

To see this more clearly, it is helpful to put it in a very general historical perspective. It has often been pointed out that the English path of capitalist development, in which capitalism developed in agriculture first, with the expropriation of the soil by landowners increasingly orientated to maximizing the returns to labour, has been rare.[5] And certainly policy-makers in the non-settler countries of Africa have mostly assumed that capitalist relations of production would be established first in the non-agricultural sector, and only later, if at all, in farming. If there has been a latent historical model, it has perhaps been that of Japan, where even today the average size of farm is under 1.5 hectares, with very little tendency towards concentration, but where productivity has risen at an average annual rate of 1.5% per annum over the last hundred years, not merely enabling the non-agricultural population to be fed largely from domestic food production, but also furnishing the original financial surpluses and initially also the bulk of the necessary foreign exchange (from silk and tea exports) for industrial investment.[6]

The 'English' route to agricultural development was deliberately closed by the colonial governments in most African colonies. Although the colonial officials of British West Africa, in particular, initially entertained hopes of developing capitalist agriculture, they soon realized that the political costs of trying to force the African population off their land and into wage labour on plantations or estates were too high; they would not be able to maintain control in face of the resistance this would generate. Instead, they consolidated the existing system of smallholdings. But, in addition, they blocked the emergence of native capitalists who might have developed capitalist firms themselves. They superintended the creaming off by foreign export houses, and later by state-owned marketing boards, of the surplus of the smallholders who produced export crops; they also resisted the entry of Africans into other fields of capital accumulation, reserving them as far as possible for foreigners.

Now, contrary to what Marx seems to have supposed when taking the English experience as his model in writing *Capital*, it is not essential to merge 'the dwarf-like property of the many into the giant property of the few' in order to get the necessary productivity increases in agriculture. Not only can very small land units be made more and more productive by

[5] On this point see the invaluable review by Gavin Williams, 'Why Is There No Agrarian Capitalism in Nigeria?', mimeo, Oxford, 1986.

[6] *Agricultural Policy in Japan* (Paris: OECD, 1974); Y. Hayami, *A Century of Agricultural Growth in Japan* (Minneapolis: University of Minnesota Press, 1975).

applying more fertilizer and using better seed, but with rising incomes and increasingly specialized outputs the farmers become increasingly dependent on purchased inputs and on personal consumption and so form a growing domestic market for industry. And, so long as there is no prospect of industry absorbing significant proportions of the rural labour force, the 'English' route of forced separation of the rural population from the land implies mass destitution. All of these are good reasons for preferring a 'Japanese' road. In the long run, as in Japan today, very small farm sizes will emerge as an obstacle to increased labour productivity because they inhibit the use of machinery. But this obstacle may be partly overcome by the co-operative use of machinery, and in any case it arises only when the scope for raising productivity by other means has been much more fully explored than it has in Africa.

But the trouble with the 'Japanese' road is that it required conditions for its success that are not easily reproduced in Africa. The Japanese state was strong and had a coherent policy for inducing the landlord class to use the surplus produced by tenant farmers for capitalist accumulation. There was also a relatively well developed rural infrastructure (especially irrigation); and in due course the state also developed the ability to provide sustained improvements in education, in agricultural research and extension, and in the rural infrastructure. In other words, it is a state-dependent road. Instead of the impetus for continuous productivity increases coming from market forces operating on capitalist farm enterprises dependent for survival on profit maximization, it must come from a state apparatus inducing millions of risk-avoiding peasant smallholders to constantly modify their farming practices, by a mixture of well-designed and well-implemented inducements and pressures, along a path of rising efficiency already mapped out by the state.

The 'English' road, in contrast, was dependent on state power mainly to clear the land of its original owners or users: thereafter, it was primarily the search for profit by increasingly market-orientated agricultural capitalists that raised productivity. In this connection, it may be noted that contrary to a popular misconception, Marx's account of 'primitive accumulation' does not assign a central role to the state in all or even most of its aspects. A careful reading of the relevant sections of *Capital*, Volume I, Section 8, will show that he saw it as largely the work of 'robber barons', who often relied on state power but for the most part did not actually employ the state apparatus. This is pertinent to what is now happening in some African countries. What appears, from the standpoint of the 'Japanese road', to be a highly dysfunctional use of state power, may be functional from the standpoint of the 'English' road; for example, an irrigation project intended to raise the productivity of smallholders may be perverted into a project which favours the emergence of large- or medium-scale capitalist farms owned by the project administrators, other bureaucrats and army officers. (This does not imply, of course, that this is a more desirable outcome; it merely draws attention to the possibility that more than one road to rural development may be implicit in what is going on.)

It is now time to try to bring together the threads of the argument. It is that the relations of production of African agriculture are such as to hold out the possibility of continued increases in productivity only if the state is effective to a degree that is rare in African conditions. It is unnecessary to postulate a generalized 'neo-patrimonialism' — a state may be relatively rational and honest and still unequal to this task. What is called for is a state capable of pursuing sustained and coherent policies in a number of related fields, above all for raising agricultural productivity. The dependence of this route on appropriate state policies is critical. The 'English' route relied much less on state policies and activities, and much more on market forces operating, after an initial use of force, on capitalist farm enterprises.

From this perspective, then, a focus on state policy-making in Africa is not inappropriate. But why should a period in which African rural output expanded so impressively have been followed by an apparently general halt, or even a contraction? Did African states formerly have coherent and appropriate policies for agriculture, but become incoherent or worse in the 1970s? This seems implausible; any plausible view must embrace a fairly wide range of factors and reflect a wide range of variations in the African experience. What follows is at most suggestive.

In the first place, a portion of the expanded output of African smallholder agriculture has always been obtained by bringing new land into cultivation. In parts of Africa such as Côte d'Ivoire, Zambia, Sudan and the Congo Republic, this is still possible, but in many others with growing land shortage (exemplified by Kenya and Zimbabwe) it is not. The onset of the drought in the early 1980s brought this into view. The implication is that further rises in output will have to come from productivity increases, which require a level of government support not previously needed, and which, in the conditions of fiscal and exchange crisis that set in from the mid-1970s (not to mention the more long-standing problems of most African states), was highly unlikely to be forthcoming.

Secondly, the most significant aspect of Africa's agricultural problem may well be the decline in food exports. The reason for this is partly that it is not easy to be sure that per capita food production has in general fallen. According to the World Bank, it fell by 12% from 1969–71 to 1980–82, but the data on which this is based are not very reliable.[7] The data on food crop exports are relatively reliable, and what they show is that, while world food exports grew in real terms by 2.7% a year from 1965 to 1970 and 5.3% a year from 1971 to 1984, Africa's share of total world trade in food fell from

[7] Sayre P. Schatz suggests (in 'Africa Food Imports and Food Production: An Erroneous Interpretation', *Journal of Modern African Studies* 24/1, 1986, pp. 177–78) that the evidence for declining per capita food production, in so far as it is inferred from rising food imports, has been misread, and in fact implies constant per capita food production over the nineteen years from 1961–63 to 1980–82.

1.5% to 0.3% over the whole twenty-five-year period. In the 1970s, the volume of African exports fell each year for virtually every agricultural commodity except sugar and tobacco.

This in turn may be explained by various factors, depending on the circumstances; for instance, in Nigeria, thanks to the oil boom, it was certainly aggravated by the relative increase of prices for crops sold on the domestic market, and the greater returns to labour to be obtained from urban employment. In many countries, as the World Bank argues, overvalued local currencies or excessive export taxes also undoubtedly hurt exports.

However, not enough attention seems to have been paid to the general effect of declining world commodity prices. This is not a recent phenomenon. Over the period 1950 to 1984 there was a real decline of over 1% a year in world agricultural prices and of 1.3% per annum in the prices of cereals, fats and oils. Some of this decline may be attributed to a tendency for rich countries to consume relatively less of these commodities as incomes rise, and to increasing supplies; but some of it seems to be due to a decline in the costs of production by advanced capitalist producers. This is evident, for example, in the case of palm oil, where plantation production in Malaysia and Indonesia used radical technical innovations to bring down costs; and in groundnuts, where the USA, following its standard agricultural pricing policies to reward high productivity, replaced Sudan, Nigeria and Senegal as the world's leading exporter.

It is true that in the ten years 1974–83 total Nigerian palm oil production rose 10% and was sold on the domestic market (population grew four times as fast); but during this same period total Malaysian production tripled (to over four times total Nigerian output) and total Indonesian production more than doubled (to exceed Nigerian output). In the same period, too, total Sudanese groundnut production was more or less constant, but yields from peasant production were falling and Sudan's survival as an exporter of groundnuts was due increasingly to its expanding capitalist sector.

These examples remind us that peasant production of commodities for export cannot be sustained unless productivity increases more or less keep up with those obtained in non-peasant production. Otherwise the peasants eventually cease to obtain an adequate return for their labour (which has become too inefficient by world standards), and their produce ceases to be exportable. In Sudan this process appears to have reached a point where more and more peasants are being forced into wage labour in the mechanized large-scale farm sector.[8] Similarly, in several West African countries, it is no longer possible for state marketing boards to set producer prices which will induce farmers to produce and still have a surplus to show when

[8] J. O'Brien, 'The Formation of the Agricultural Labour Force in Sudan', *Review of African Political Economy* 26, 1983; and A. El-Mekki, 'Peasants and Capital: The Political Economy of Oilseeds Marketing in the Sudan', PhD thesis, University of Toronto, 1985.

the crops are finally disposed of on the world market. The gap that used to exist between the world price and a viable producer price has become too narrow.

In the case of some crops it is perhaps impossible for world levels of efficiency to be matched within the social relations of household production (this seems likely to be true for palm oil, for example), but many crops can, theoretically, be produced efficiently on smallholdings. The trouble is, once again, that the state policies needed to secure this are not likely to be forthcoming in many African countries. On the contrary, what is more characteristic in many situations is the use of state power by emerging native capitalist classes to acquire land for capitalist farming, as has happened in Nigeria, northern Ghana, Sudan, parts of Kenya and elsewhere. It is true that capitalist farming in Africa has generally secured state support not only for the initial appropriation of land, but also in the form of subsidized inputs, price advantages and the like, raising still further the social costs of this 'road'; and it cannot be sustained unless there is an adequate labour force with no real alternative to wage-labour. But it seems important to acknowledge that it has, in some countries, been a way in which advances in agricultural production have been achieved, and may be so again in others in the future.

In practice, the pattern seems likely to be mixed, within countries as well as between them. Regions of capitalist farming coexist alongside large regions of simple commodity production in many countries (e.g. Zimbabwe, Zambia, Kenya, Côte d'Ivoire, the Sudan), while the development of various forms of integration between household farms and the corporate purchasers and processors of their crops (irrigation schemes, outgrower schemes, contract farming, etc.) constitutes a sort of intersection between the two 'roads' which is quite significant in some countries (Côte d'Ivoire, Kenya and Sudan, among others). And in spite of resistance to the idea among protagonists of 'the peasant', a process of internal differentiation is inevitably taking place which is, however gradually, leading to the emergence of some small-scale capitalist farms, on the one hand, and increasingly wage-dependent labourers on the other, within many predominantly 'household' farming areas.

Already in the early 1960s Samir Amin was calling for the destruction of the African family farm and the individualization of holdings, on the grounds that productivity could never be raised within the framework of the traditional household unit.[9] Today he would no doubt qualify this, but the basic issue remains: a functional equivalent of the impact of 'primitive accumulation' on African agriculture must be found — that is, some means for (i) permitting and (ii) forcing it to achieve continuous improvements in

[9] 'Bureaucracy will not justify itself historically as a progressive force in West Africa or anywhere else except to the extent that it succeeds in completely transforming the economic base of society to realise primitive accumulation' (Samir Amin, *The Class Struggle in Africa*, Reprint No. 2, Africa Research Group (Dar es Salaam: University of Dar es Salaam, n.d.)).

productivity. This does not exclude socialist means, if social forces capable of sustaining them can be mobilized. The only alternative, apart from the expansion of capitalist farming, seems to be a growing risk of some new form of colonization, from chronic dependence on food aid and/or current budgetary support from abroad, to the role of economic and political satellites such as the apartheid regime in South Africa envisaged for its neighbours.

7

Learning from the Kenya Debate

What was the 'Kenya debate' about? Ultimately, it was about whether capitalistic development is possible at the 'periphery' of the world capitalist system, or, more accurately (since in the real world nothing is excluded absolutely), 'whether or not there are theoretical reasons for thinking that the ex-colonies cannot "adopt the bourgeois mode of production" and develop their productive forces within it' (in the words of the paper which originally provoked the debate).[1] In the late 1970s, English-speaking students of African development, influenced by the rather categorical versions of dependency theory popularized by development theorists such as Andre Gunder Frank and Samir Amin, were largely convinced that Africa could not develop under capitalism. Consequently, questioning this in relation to Kenya provoked a general debate focused on an African country which had been more intensively researched than most. The aim of this chapter is to look back at the debate in the light of what has happened since — both in development thinking and in Kenya. It begins with a very brief summary of the debate: readers who are familiar with it may well prefer to start at the following section.

The 'Kenya debate'

Dependency theory, which had informed my previous work on Kenya, suggested that there were three seemingly insuperable obstacles to the development of Kenya under capitalism: (i) the structures of the colonial economy, which appeared unconducive to the efficient use of national resources but which were taken over largely intact by the incoming independent regime; (ii) the monopoly power of foreign capital, on which the

*First published in David E. Apter and Carl G. Rosberg (eds), *Political Development and the New Realism in Sub-Saharan Africa* (Charlottesville: University Press of Virginia, 1994).

[1] Colin Leys, 'Capital Accumulation, Class Formation, and Dependency: The Significance of the Kenyan Case', *Socialist Register* (London), 1978, p. 241.

small African business class seemed heavily dependent; and (iii) the state, which seemed increasingly preoccupied with the use of economic resources (access to credit, import licences, profitable franchises, etc.) to buttress the political power (and the personal fortunes) of the president and his closest associates.[2]

But by 1978 I came to think that the question of the further development of Kenya under capitalism should be regarded as open, not foreclosed, for two main reasons: first, Michael Cowen had shown that before colonial rule there had been a capital-accumulating class among the Kikuyu of Central Province and that this class, after many vicissitudes under colonialism, had emerged as a core element in the class that had come to dominate Kenyan politics since independence; second, it was evident that this class had succeeded in occupying one sector after another of the modern Kenyan economy during the post-independence years of 1964–76 to the point where they were, in 1977, even beginning to enter the manufacturing sector, and that, while they were doing all this, not only was there no catastrophic drop in output (as had happened under Amin in Uganda, for example), but the average annual real growth rate was 6%. It was also clear that their access to state power had been an important instrument of their advance; through legislation and administrative action they had been able largely to take over large-scale capitalist agriculture and to become established in trade, tourism, urban and coastal real estate, small-scale building, import–export trade and so on. There was also evidence of growing social integration among the members of this class, and there were signs of the emergence of a consciousness of their class position and interests — the beginnings of an awareness of the requirements (legal, international, etc.) for the consolidation and defence of these interests — including an emerging critique of the arbitrary and rapacious behaviour that had been indulged in by leading members of the Kenyatta regime.

Looking back at Marx's analysis of why capitalism first became established when and where it did (i.e. in the seventeenth century and in Britain, as opposed to, say, the Netherlands), I noted that he had explained it (in *Das Kapital*, Volume 1; Part 8) in terms of a combination of factors that happened to occur then and there, and not at some other time or place; and I argued that, when looked at in this way, Kenya might likewise be seen as exceptionally equipped (by African standards) with the combination of factors (in particular, its exceptionally strong indigenous accumulating class) needed to overcome the contemporary constraints to capitalist development to which dependency theory pointed.

The debate was joined when two experts on industrialization in Kenya, Raphael Kaplinsky and Steven Langdon, argued that this was so improbable as to be virtually impossible. Kaplinsky took the view that any peripheral country that does not have either a large internal market or the good

[2] Colin Leys, *Underdevelopment in Kenya: The Political Economy of Neo-Colonialism, 1964–1971* (Berkeley: University of California Press; London: James Currey, 1975).

fortune to be able to profit from a particular and perhaps unique phase in the global development of capitalism (like the newly industrialized countries (NICs)) faces absolute limits to capitalistic accumulation, whether or not it has a domestic class of accumulators with precolonial roots: 'The inbuilt contradictions of economies of this type [i.e. other than those with large internal markets like India or Brazil] make it difficult to foresee that such a pattern of accumulation — with or without foreign capital — can proceed in a viable form.'[3] Langdon's position was similar, in that he considered that 'the Kenyan bourgeoisie's alliance with foreign capital...means that the indigenous bourgeoisie cannot carry forward a broad transformation of social relations...because it is dependent on foreign capital', whose technology choices constituted, in his eyes, an insuperable barrier to such a transformation. He differed from Kaplinsky in explicitly not ruling out the possibility of a capitalist transformation of Kenya ('this is not to say that [a broad capitalist transformation of social relations] is...impossible in Kenya'), but he could imagine it occurring only if for some as-yet-unforeseen reason this suited the needs of foreign capital.[4] The domestic capitalist class, in his view, was dependent on foreign capital and, by virtue of that dependency, incapable of bringing about such a transformation.[5]

Why return to the Kenya debate?

In general, such debates are not resolved but tend to peter out in mutual fatigue and indifference, while the central issues that seemed to be so importantly at stake cease to compel attention and are replaced by new preoccupations. Noting this in his 1985 review of the debate, Gavin Kitching

[3] Raphael Kaplinsky, 'Capitalist Accumulation in the Periphery: The Kenyan Case Re-examined', *Review of African Political Economy* 17, January–April 1980, pp. 103–4.

[4] Steven Langdon, 'Industry and Capitalism in Kenya: Contributions to a Debate', in Paul M. Lubeck (ed.), *The African Bourgeoisie: Capitalist Development in Nigeria, Kenya, and the Ivory Coast* (Boulder: Lynne Rienner, 1987), p. 346. Langdon's paper was originally presented at a conference in 1980 and was commented on, together with Kaplinsky's, by Gavin Kitching in his 'Politics, Method, and Evidence in the "Kenya Debate"', in H. Bernstein and B. Campbell (eds), *Contradictions of Accumulation in Africa: Studies in Economy and State* (Beverly Hills, California: Sage Publications, 1985), pp. 115–52.

[5] There were several other contributors to the debate, notably Nicola Swainson, 'The Rise of a National Bourgeoisie in Kenya', *Review of African Political Economy* 8, 1977, pp. 39–55, and *Foreign Corporations and Economic Growth in Kenya* (London: Heinemann, 1980); J.S. Henley, 'Capitalist Accumulation in Kenya: Straw Men Rule OK?', *Review of African Political Economy* 17, January–April 1980, pp. 105–8; and Colin Leys, 'Kenya: What Does "Dependency" Explain?', *Review of African Political Economy* 17, January–April 1980, pp. 108–13. All of these contributors broadly took the line reflected in my 1978 paper. Björn Beckman, in 'Imperialism and Capitalist Transformation: Critique of a Kenyan Debate', *Review of African Political Economy* 19, September–December 1980, pp. 48–62, took a line closer to the one I now hold; see also his 'Imperialism and the "National Bourgeoisie"', *Review of African Political Economy* 22, 1981, pp. 5–19. Kitching, 'Politics, Method, and Evidence', op. cit., adopted a position close to that of Warren (discussed below).

suggested that in this instance it was because all the contending participants were unclear (or, if clear, uncomfortable) about the political implications of their respective theoretical positions. Being unwilling to confront these explicitly, they (we, that is) expended energy on disputing the significance of the factual evidence on Kenya's post-independence development, the nature of the manufacturing fraction of the indigenous bourgeoisie, the role of the state and so on, rather than acknowledging that what the debate really brought into focus was the very different and problematic alternative political commitments and strategies that we were implicitly endorsing.

According to Kitching, those on the Cowen–Leys–Swainson side of the debate really believed that a capitalist development was the best hope currently available for countries like Kenya. Kaplinsky and Langdon, for their part, would have liked such development to be possible (Langdon, in fact, explicitly speculated on the possibility that its necessarily limited life might be prolonged by a strategy of income redistribution — enlarging internal markets for manufactures — introduced by a populist coalition of enlightened state technocrats and what Latin American analysts like to call the 'popular sectors'); but they actually believed it to be impossible, although they did not spell out the political implications of their pessimism. Kitching himself looked for the construction of 'a sophisticated socialist movement ... among the working classes of Africa as they slowly expand through time'. He considered that this would 'take a long time and that a prolonged period of struggle against a developing capitalism is one of the prerequisites of its creation. To that extent I am "happy" to see continued capitalist development in black Africa (though this may or may not occur).'[6]

None of these political stances would, of course, have been very acceptable to progressive Kenyans, and it was an unstated (and perhaps unacknowledged) awareness of this, Kitching thought, that led to the debate being abandoned by all the original participants, who on the whole liked to think of themselves as progressive too.

Whether or not Kitching's diagnosis of the reasons for the inconclusive outcome of the debate was correct, the impulse behind his intervention was sound: the failure to resolve such debates means that the confusions involved are likely to recur, after some time, in new forms. While we must accept that in social life problems recur constantly and that this will often be little affected by the extent to which social scientists succeed in identifying and understanding them, this is not an excuse for social scientists not at least trying to reach agreement about what the problems are and how they should be understood. In this case, where we are dealing with issues that have very practical current political implications — the development strategies of governments and aid donors, the political strategies of political and social movements — the obligation to try to draw up the accounts of the debate is surely real.

[6] Kitching, 'Politics, Method, and Evidence', op. cit., pp. 147–48.

For, in its widest sense, the debate was a confrontation between competing paradigms or metatheories about what is happening in the ex-colonial world and our reasons for studying it. In particular, it was an encounter between the dependency approach to development, and that approach's implicit political commitments, and the approach of 'orthodox' or 'classical' Marxism, and its implicit commitments. And, since, as Dudley Seers and others have pointed out, classical Marxism has much in common with classical political economy, from which modernization theory is also derived, the Kenya debate can be read as a text in which, to varying degrees, all the principal attitudes, assumptions and commitments current in development studies today are manifested and tested.[7]

This chapter does not try to offer a full reading of the Kenya debate considered in this way. It aims simply to consider the theoretical context in which the debate occurred and offers some conclusions about the relation between the dependency and classical Marxist approaches that the passage of time suggests are reasonable. Then it returns to the debate in the light of what happened in Kenya in the 1980s, indicating some empirical questions which, with hindsight, it might have been more useful to ask.

The theoretical context

As between dependency theory and 'classical' Marxism, and perhaps 'modernization' theory too, there is at least general agreement that the ex-colonies (or 'underdeveloped', 'Third World', or 'periphery' countries) exist in a world in which there is a world market and increasingly a world system of industrial production, dominated by the economies and state power of the advanced capitalist countries.[8] Roughly, dependency theorists tend to think

[7] Dudley Seers, 'The Congruence of Marxism and Other Neo-classical Doctrines', Discussion Paper, Institute of Development Studies, University of Sussex, August 1978; Alain Lipietz, 'Marx or Rostow?', *New Left Review* 132, March–April 1982, pp. 48–58. I do not agree with the polemical thrust or with all of the analysis of Lipietz's critique, but the general point — that the political implications of a modernization and an orthodox Marxist interpretation of development in the Third World may be hard to distinguish in practice — is valid. See also Manfred Bienefeld, 'Dependency Theory and the Political Economy of Africa's Crisis', *Review of African Political Economy* 43, 1988, pp. 68–87, especially pp. 84–85.

[8] The label 'dependency theory' is used here to cover all the variants of this approach, from the nineteenth-century Latin American historiography relied on by Andre Gunder Frank in his *Capitalism and Underdevelopment in Latin America* (New York: Monthly Review Press, 1967) and the very general theory advanced in that book, to the much more tentative theoretical perspectives offered by Henrique Cardoso and Enzo Faletto in *Dependency and Development in Latin America* (Berkeley: University of California Press, 1979), and all the many middle-level theoretical efforts and monographs belonging to this school of thought. The label 'classical Marxism' is used to cover the point of view of writers like Geoffrey Kay (in *Development and Underdevelopment: A Marxist Analysis* (London: Macmillan, 1975)) and Michael P. Cowen and Bill Warren, who see themselves as developing rather than revising the analysis of capitalism proposed by Marx in *Das Kapital* and Lenin in *The Development of Capitalism in Russia*.

that it is not possible for the ex-colonies to overcome their poverty and technical backwardness by trading in this world market or by adopting the mode of production of the advanced countries, based on private property in the means of production, because the prior dominance of the advanced countries in both the market and the global division of labour means that it will never bring them the desired benefits. Classical Marxists, on the other hand, tend to think that (i) the geographical spread of capitalist production relations has been going on since at least the sixteenth century and has been accelerating since the end of formal empires; and (ii) the 'law of value' (in the common-sense meaning of competition in world markets from goods produced ever more cheaply through the constantly rising productivity of labour that is the distinctive consequence of capitalistic production) gradually forces all countries to adopt this mode of production, on pain of being pauperized if they don't. As Marx put it, cheap manufactures eventually 'batter down all Chinese walls' that hitherto protected the precapitalist periphery.

The extreme positions of the two approaches are represented by the early writings of Frank, on the one hand, and by Bill Warren's posthumous book, *Imperialism: Pioneer of Capitalism*,[9] on the other. Frank originally said that capitalist growth was impossible for the periphery so long as it was linked to the metropoles and he advocated a revolution capable of leading an autonomous growth strategy based on cutting those links of dependency. Warren at least implied that capitalist growth was inevitable, apparently everywhere. In practice, few people today hold these extreme positions. Dependency theorists like Fernando Enrique Cardoso, noting that Brazil was undergoing rapid growth and a rapid transformation of precapitalist social relations in the 1960s, argued that dependency relations only make the course of such 'dependent accumulation' more difficult and its long-term results more problematic, less certain to be beneficial in the way the modernization school assumed. Classical Marxists like Geoffrey Kay or John Sender and Sheila Smith, on the other hand, while arguing that the spread of capitalist market and production relations is more or less universal, do not maintain that it necessarily results everywhere in sustained economic growth, let alone eventual 'high mass-consumption society' (in Walt Rostow's famous phrase).[10]

At first sight, then, the differences between the two approaches would seem to have become relatively narrow. Why should *dependentistas* not agree that the spread of capitalism is indeed more or less universal and the classical Marxists not agree that dependency theory points out many of

[9] See Frank, *Capitalism and Underdevelopment in Latin America*, op. cit., and 'The Development of Underdevelopment', in his *Latin America: Underdevelopment or Revolution* (New York: Monthly Review Press, 1969); Bill Warren, *Imperialism: Pioneer of Capitalism* (London: New Left Books, 1980).

[10] Cardoso and Faletto, *Dependency and Development in Latin America*, op. cit.; Kay, *Development and Underdevelopment*, op. cit.; John Sender and Sheila Smith, *The Development of Capitalism in Africa* (London: Methuen, 1986).

the ways in which, under contemporary conditions, the results of this process are unlikely to bring to at least very many underdeveloped countries the sort of benefits that people have in mind when they talk about 'development'?

There are no doubt many reasons why such a convergence has not occurred, from intellectual inertia to political and personal factors of all sorts. But, at the level of academic exchange, one can see that there are some latent assumptions that still divide the two schools. *Dependencistas* tend to minimize the significance of cases such as Brazil or India (large internal markets) or the smaller NICs (exceptional international conjuncture in the 1960s and 1970s) and to go on thinking that, for virtually all the remaining underdeveloped countries, growth under capitalism, let alone any beneficial kind of development, really is impossible. (This was clearly the view of Kaplinsky and Langdon in the Kenya debate.) On the other hand, classical Marxists tend to believe that, although the process of transition to capitalism at the periphery will not be pleasant (i.e. it will be uneven, full of crises and contradictions, and will cause widespread human suffering, just as it did in Europe), it nevertheless will eventually occur, and it will replace the 'idiocy of rural life' with industrialization, without which development is meaningless.

Classical Marxists point out that the *dependencistas* seem to assume the possibility of an alternative historical path of development which the periphery might follow — free from the irrationality, unevenness and suffering involved in capitalist development — whereas no such path is shown to be possible, even in theory, let alone in current reality. Like Mrs Thatcher in 1980, the classical Marxists declare that for the time being 'there is no alternative' to capitalist development. Yet *dependencistas* might point out with equal justice that the classical Marxists seem to assume that this development will occur in such a way that, whatever its costs, it can be relied on eventually to industrialize the whole periphery.[11] But, just as capitalism developed unevenly in today's industrial economies, marginalizing and pauperizing whole regions while concentrating production and wealth in others, so on a world scale whole countries and even regions seem likely to become chronically, and for all practical purposes permanently, marginal, with the added problem that their labour power is not even required elsewhere but on the contrary is excluded by increasingly rigorous controls over migration, while rates of population growth undreamed of when today's industrial countries were developing are pushing them rapidly toward terminal Malthusian crises.[12]

[11] This point is developed in relation to the debate about development in the Atlantic provinces of Canada in Colin Leys, 'Populism, Socialism, and the Struggle of Primary Producers', in Bryant Fairley, Colin Leys and James Sacouman (eds) *Restructuring and Resistance: Perspectives from Atlantic Canada* (Toronto: Garamond, 1990).

[12] The potential of the acquired immune deficiency syndrome (AIDS) epidemic to anticipate such crises in several African economies has, it seems, still to be registered theoretically in the wider debate about African development.

Thus both sides make implausible assumptions, and neither has a tenable political position — at least for most of sub-Saharan Africa. *Dependencistas* tend to assume an alternative historical path which is not plausibly shown to exist even in theory, let alone one which existing organized social forces appear capable of implementing. Classical Marxists and modernizers, on the other hand, tend to assume a future outcome which is very hard to imagine for many countries, and which in any case seems to postpone practical action on behalf of the 'popular sectors' far into that improbable future.

Of the two, the *dependencistas*' implausible assumptions are the less incapacitating, and no doubt this is one of the reasons why the dependency perspective became (as Björn Beckman pointed out) broadly dominant among progressive thinkers and activists in Africa and — for some time, at least — among the so-called development community in the advanced capitalist countries, including academic students of development.[13] Dependency theory focuses on the 'bad side' of the capitalist development that is occurring in the periphery, and it identifies the chief beneficiaries of the process (foreign capital, local compradors) as agents of the ills it brings. It rarely confronts the practical question of what social forces will support the alternative development strategy that it implies is available, and how their apparently conflicting interests will be reconciled within it.[14] So it gives the appearance of being compatible with the populist–nationalist sentiments of most of those who contest the prevailing direction of development inside the periphery countries. But this does not make the dependency position intellectually or politically acceptable. Evidently both sides need to abandon their untenable assumptions and converge around a programme of enquiry in which both the long-term economic implications of what is happening in any given country and the corresponding implications for political action and alternative development strategies are seen as much more open. It would also be healthy for the political and strategic issues to be discussed explicitly, not least by the 'aid community', who have great influence in the countries concerned, while not sharing their fate. The shared assumption should be that the periphery countries are indeed being forced to 'adopt the bourgeois mode of production', and the shared question should be: With what consequences for each one? From the dependency perspective we can draw a set of questions about the contradictions involved, which tend to restrict the speed of the adoption process, determine its specific forms (e.g. 'dependent accumulation') and define its potential costs and benefits for various segments of the population. From classical Marxism we get a focus on the forces (economic, political and cultural) that tend to

[13] Beckman, 'Imperialism and Capitalist Transformation', op. cit., p. 60.

[14] For an interesting exception, see Björn Beckman, 'Peasants and Democratic Struggles in Nigeria', paper presented to the AKUT conference on Labour and Democracy, Uppsala, September 1986. The difficulties that Beckman bravely confronted in this paper are highly significant.

sustain the process and, in particular, the new social forces created by it, whose struggles will be, if not decisive, at least very important in deciding how, and with what consequences, the contradictions are resolved.

But neither approach warrants any determinism. The questions resulting from combining them should be looked upon as genuinely open. Many of today's industrialized countries once exhibited many of the characteristics that are seen as dooming today's peripheral economies to perpetual stagnation.[15] The internal contradictions in any given individual country today may seem more acute, and the external environment of late twentieth-century global capitalism, dominated by multinational corporations (MNCs), regulated by the International Monetary Fund (IMF) and policed by the USA, may seem more hostile, than was the case one or two hundred years ago for any of today's industrial countries. It is also true that not every country can be a pole of capital accumulation. But the fact is that we cannot know in advance how the contradictions will develop or be dealt with in any given case, or how the social forces at work inside and outside a country will develop and interact — least of all in the post-USSR, post-Warsaw Pact age. There are certainly more ways in which contradictions will be resolved, and more possibilities of social evolution within the complexities of the evolving world market, the world system of production and the world political order, than have yet been seen or even imagined. It is impoverishing, as well as dogmatic, to suppose that we can have in general a 'correct' political line in relation to the peripheral countries.

Rather, the time has come to acknowledge that only the analysis of the social forces at work in each individual country and their political and organizational capacities can permit intelligent political choices to be made or alternative development strategies to be evolved. In some cases, it may make sense to adopt a 'proletarian' stance, looking to the political potential of the organized working class as a force for social and economic advance; in others, this may make no sense. At any rate, a discussion of the sense it makes ought to be as much part of the agenda of development research as the study of any other aspect of social change.

The relevant facts

What sort of facts are relevant to the issues raised in the Kenya debate? What principally strikes one, in the light of such reflections, is that in the Kenya debate attention was focused on the wrong factual issues. Critics have argued that the empirical evidence adduced was often ambivalent, capable of being read as supporting either position: a question of whether

[15] See Gavin Kitching, 'The Role of a National Bourgeoisie in the Current Phase of Capitalist Development: Some Reflections', in Lubeck, *The African Bourgeoisie*, op. cit., pp. 27–55.

the bottle was seen as half-empty or half-full.[16] There is undoubtedly some force in this criticism, but the more serious defect was that the debate focused on the wrong sort of things.

Specifically, Kaplinsky and Langdon focused on five main issues: (i) the capability of the Kenyan capitalist class, and specifically the extent to which it was displacing foreign capital; (ii) whether the Kenyan state was under its control, or at any rate responsive to its wishes, *vis-à-vis* foreign capital; (iii) whether capitalist production relations had spread since independence; (iv) whether the growth that had occurred was 'successful' or 'viable'; and (v) whether such growth was likely in the future.

These issues reflected their understanding of what had to be denied if the dependency interpretation of Kenya was to be upheld, and some of the confusions involved have been sufficiently pointed out by Björn Beckman and Gavin Kitching. The last two of the five issues addressed by Kaplinsky and Langdon are self-evidently unrewarding, since growth that seems unsuccessful from a dependency viewpoint obviously can seem significant and potent from an opposite viewpoint. The question of how far capitalist relations of production had actually spread was more interesting but was equally liable to degenerate into a mere contraposition of two incompatible but equally plausible ways of reading the same evidence.[17] But it is the issue of the capacity of the domestic bourgeoisie and the closely related issue of the role of the state that best illustrate the way in which abandoning the false opposition of dependency theory and classical Marxism allows us to go beyond the inconclusive results of the Kenya debate.

Kaplinsky and Langdon read the new antidependency stress on the growth of an internal capitalist class as reflecting the view that the existence of such a class was almost a sufficient condition for capitalist growth. Hence,

[16] See Beckman, 'Imperialism and Capitalist Transformation', op. cit., pp. 53–54, and Kitching, 'Politics, Method, and Evidence', op. cit., pp. 127–31. See also Martin Godfrey, 'Kenya: African Capitalism or Simple Dependency?' in Manfred Bienefeld and Martin Godfrey (eds), *The Struggle for Development: National Strategies in an International Context* (London: Wiley, 1982), pp. 265–91, especially pp. 272–74.

[17] Kitching took Langdon to task for using wage employment as a proportion of total employment as an indicator of the spread of capitalist production relations, arguing that the 'development of commodity production under capitalism' need not involve the 'universalisation of the capital/wage labour relationship, but was registered in a long-term increase in commodity output and labour productivity, and so on' ('Politics, Method, and Evidence', op. cit., p. 124). While this is true, I think Langdon was also justified in pointing out that if wage-labour remains constant, at a small proportion of total labour, it raises a question about the viability of capitalist development — about the ability of capital to make total labour more productive and to accumulate surplus value from it, which is the necessary basis of capitalist growth. On the other hand, it is clear that the scale of non-enumerated wage-labour is now of the same order as enumerated wage-labour, and, since by definition its precise size is unknown, it is not possible to say whether trends in wage employment in Kenya show that capitalist production relations are 'spreading' or not. The figures used by Langdon to illustrate his point were for enumerated wage-labour only (Langdon, 'Industry and Capitalism', op. cit., pp. 352–53).

to defend the dependency position, they had to show that this class was neither efficient nor strong, nor independent of foreign capital (let alone in the process of displacing it), and in any case no 'better' than foreign capital in terms of its use of local inputs, levels of protection, efficiency, etc. In the same way they assumed that the state must be shown not to be an 'instrument' of this class. By now it is easy to see that the existence of an internal capitalist class, even a relatively nationalist and efficient one, is not a sufficient condition for capitalist growth; in some circumstances perhaps it might not even be a necessary condition. Similarly with the state: whether or not it is in general more responsive to the domestic capitalist class than to foreign capital is a priori not conclusive for the prospects for capitalist development either. The whole question was wrongly posed in these terms. What needed to be asked was what were the chief issues affecting the prospects for capitalist growth, and what dynamic forces were at work that would most likely determine how these issues would be resolved.

None the less, it seems perverse to suppose that the prospects for capitalist development are not affected in any way by the size and nature of the domestic capitalist class; there was a rational kernel in the emphasis laid on it in my 1978 article, even if the reasons were unclear. First, foreign manufacturing capital, organized in MNCs, is not interested in undertaking or organizing the production of most of its needed inputs, let alone local production in every country where it is established. To establish the linkages that are necessary to the deepening and extension of capitalist production, an internal capitalist class is usually necessary, and the fact that by 1977 Kenyan capitalists had taken over substantial parts of some of the most important sectors of the Kenyan economy without reducing the overall rate of growth suggested that they had the capacity to make these linkages.

Second, a politically powerful domestic class of capital with an orientation to production is likely to be necessary to secure the political (as opposed to the economic) conditions for the expansion of capitalist production. Otherwise the state is unlikely to sustain the policies needed to resolve the multiple contradictions involved in the process of expanding capitalist relations of production and sustaining capital accumulation in face of opposition from other classes and interests. The evidence that suggested such a class might be forming in Kenya, with a productivist culture stemming from the precolonial history of the Kikuyu, was the second reason for the emphasis on it in my 1978 article. In retrospect, it seems to me that the second of these reasons was the more important. It now seems to me less important that the domestic bourgeoisie should be efficient — technically, financially or otherwise — as capitalists, as individual accumulators, than that they should be competent politically, as a class: that they should, as a class, recognize the requirements of capital accumulation for capital as a whole and be able to see to it that these requirements are met. This is the issue that should have been the main empirical focus of the Kenya debate.

This is not to say that the question of the technical competence of the domestic capitalist class is unimportant. A comparison of the fate of retail trade after its Africanization in Kenya with the results of the same experience in Uganda or Tanzania might be instructive in this respect. There is also a significant connection between this question and that of the class's political competence, in that the African bourgeoisie in Kenya has chosen to accept the continued presence of most of the Kenyan-born Asian capitalists, whose technical competence has been a major factor in maintaining efficiency and output during the period of African entry into the accumulation process.[18] But this very point reinforces the prior importance of the domestic class's political role, since technical competence is — within limits, to be sure — something that can be acquired by various means, provided the need for it is recognized. What cannot be imported or purchased are the cohesion, organization, leadership and political skills needed to secure the hegemony of the accumulating and above all the productive class, so that state policy serves the needs of capitalist production.

If this is accepted, the interesting empirical questions to ask about Kenyan development in the years since independence are different from the ones that occupied most of the space in the Kenya debate. Much of the macroeconomic evidence remains as capable as ever of conflicting interpretations. Thus real gross domestic product (GDP) rose at an average rate of 4.1% a year from 1980 to 1989, keeping pace with population growth; this was barely two-thirds of the average rate for the first fifteen years of independence but still contrasted very favourably with the rest of sub-Saharan Africa, which as a whole grew at an average annual rate of only 2.1% over the same period. On the other hand, the structure of the economy did not change during the 1980s, agricultural productivity stagnated, the balance-of-payments gap grew, foreign exchange reserves fell dramatically and the external debt burden rose from 50% of gross national product (GNP) in 1980 to 72% in 1989 (although this too still compared favourably with the 1989 average of 97% for sub-Saharan Africa as a whole). There were thus serious failures, as well as achievements. The question that most needs to be asked is: How far has the class that has the greatest interest in surmounting and resolving the problems confronting capitalist development in Kenya identified these problems or shown itself able to tackle them?

Capitalist development in Kenya

A good starting point is the analyses made on behalf of British investors

[18] The significance of this aspect of Kenyan development has been the subject of some important research by David Himbara. Himbara has argued that the Kenyan 'Asians' have actually been wholly, not partly, responsible for the domestic element in successful accumulation in Kenya: see David Himbara, *Kenyan Capitalists, The State, and Development* (Boulder, Colorado: Lynne Rienner, 1994).

by papers like the *Financial Times* or organizations like the Economist Intelligence Unit. In the mid-1980s British capital had some £1 billion invested in Kenya (far ahead of any other country; US companies, the next largest foreign group, had some £325 million).[19] This literature immediately establishes two things. First, there is a broadly shared understanding of the main problems facing capitalist development in Kenya. Second, the 'aid community' functions, from the point of view of foreign capital, very much as a sort of state apparatus of international or global capital, supplementing the technical capacities of the peripheral state and reflecting the concerns of the 'international' (i.e. British, US, German, etc.) corporate bourgeoisie. This is obvious enough in the case of the institutions designed for this purpose, such as the World Bank, the IMF, the International Development Agency (IDA), the Organization for Economic Co-operation and Development (OECD), and the European Community; but bilateral aid to Kenya is also co-ordinated internationally, via a 'consultative group' of bilateral donors meeting in Europe with Kenyan representatives under the sponsorship of the World Bank. The role of Kenyan capitalists and the Kenyan state may be important, but by now they are operating as part of a wider and very powerful system for managing capital accumulation on a world scale. The way the Kenyan state (the Kenyan 'authorities' in aid community parlance) grapples with the contradictions of Kenyan development is strongly influenced by the general consensus of global capital about what is necessary for capitalist development in Kenya.[20]

This consensus is not necessarily wise, nor is it always complete. For example, in 1982–85 the World Bank was criticized within the aid community for a dogmatic concern with liberalization measures (e.g. in relation to maize marketing) not based on real considerations of efficiency.[21] Peter Coughlin has similarly shown that uncritical, dogmatic pressures from the IMF and the World Bank for import liberalization in the mid-1980s, supposedly in the interest of forcing manufacturers to become efficient and switch to export markets, failed to respect the time it takes to establish and deepen efficient manufacturing sectors or to recognize the extent to which their inefficiencies were due to factors beyond the manufacturers' control. In general, he argued persuasively, the World Bank and the IMF failed to recognize the extent to which import substitution in Kenya, far from being exhausted, 'had just begun' and would remain an essential counterpart to

[19] *Africa Research Bulletin*, 30 June 1986.

[20] Incidentally, it is hard to imagine that anyone reading this literature could suppose that international capital has an interest in keeping Kenya 'underdeveloped', so that its weight in the Kenyan economy made successful capitalist development there less likely, as some early dependency literature used to suggest.

[21] Martin Godfrey, *Kenya to 1990*, Economist Intelligence Unit Special Report No. 1052 (London: The Economist, 1986), pp. 23–24.

export-based industrialization for a long time to come.[22]

However, in general one can see that the aid community, with the leverage of structural adjustment loans conditionality, has played a central and largely consistent role, politically as well as technically, in identifying and attempting to resolve the contradictions of capitalist development in Kenya since the 1970s. Martin Godfrey, after surveying the successive policy conditions imposed by the aid community down to 1985, concluded that 'a negative interest rate, overvalued currency, "high" wage economy, has been transformed into a positive real interest rate, market exchange rate, low wage economy, in which agricultural producers continue to be offered attractive price incentives...it clearly reveals the influence of the IMF and the Bank'.[23] What did this community see as the major problem, then, confronting further development in Kenya?

The 'fundamental' or 'underlying' problem was seen as population growth, which was believed to be the highest in the world, at 4.1%.[24] There will be an additional workforce of several millions by the year 2000, most of whom cannot expect to be employed for wages and who will increasingly lack access to land. Without very costly irrigation, only some 20% of the country is arable. In addition, land is still the most reliable source of capital accumulation by the Kenyan African bourgeoisie, as well as a source of security, social status and political influence. Consequently land reform, which in theory could provide most of the new workforce with a livelihood for at least some time to come, reducing the risk of political instability and expanding the domestic market for manufactures, is not discussed. Diana Hunt, in particular, has argued that

> a radical redistribution of land in both the large and small farm sectors, with a 3-hectare ceiling on existing farms in the high potential areas (larger in the low potential areas) might...generate some 3.1 million new farms and some 1.55 million man-years of employment...There can be no doubt that land reform presents the single most effective means available of expanding productive employment in agriculture within this period [i.e. to the year 2000]...[But] so far the landed elite among those who hold power in the Moi regime have successfully prevented any moves in this direction. This problem cannot change unless a majority of those landowners come to see personal as well as social advantage

[22] Peter Coughlin, 'Toward a New Industrialisation Strategy in Kenya?', Discussion Paper, Industrial Research Project, University of Nairobi, 1987, p. 2. Economic liberals would, no doubt, argue that, even if the source of inefficiencies lies beyond the control of the manufacturers, they remain inefficient none the less and should cease production in the interest of the whole economy. Fortunately, the heyday of this credo seems to be passing and a more realistic view, concerned with the building of long-term productive capacities, seems to be coming back into vogue.

[23] Godfrey, *Kenya to 1990*, op. cit., p. 18.

[24] Interestingly, no mention ever seemed to be made of the probable impact of AIDS on the World Bank's projection of 35 million Kenyans by the year 2000, or 80 million by 2035. In Uganda rates of human immunodeficiency virus (HIV) infection of 15% have been found in surveys of rural areas, with much higher rates in towns. Informed opinion holds that similar rates will be found in Kenya after a lag of several years.

> in such measures. For them to do this would mean that they must elect to forgo the main and most secure outlet for capital accumulation that has until now been available to them.[25]

This puts increased pressure on the state to clear the way for more rapid industrialization. The easy phase of import-substitution industrialization (ISI), if not ISI in general, is considered to be over, so that the strategy has shifted to making manufacturing efficient enough to sell in export markets. The central idea here is that Kenya must emulate the NICs and compete in world markets, at least in those most geographically accessible. There are many doubts as to whether this is practicable, even if the government can stand the strain of enforcing the very unpopular further structural adjustment measures that the aid community sees as necessary for the purpose. As Godfrey frankly expressed it:

> Is the growing pressure of labour on land, combined with the recession and the IMF/World Bank stabilization and structural adjustment programme, transforming Kenya into a land of cheap labour, expensive capital and competitively priced currency, thus poised to become an exporter of labour-intensive manufactured goods? Kenyan wages have still some way to fall before they are as low as those in some South and South East Asian labour surplus economies.'[26]

But the need to expand manufacturing's share of GNP, mainly via an expansion of manufactured exports, is seen as inescapable (with accompanying measures to try to keep agricultural output rising as fast as population and to expand earnings from tourism). What, then, are seen as the main requirements for and obstacles to this?

Here the answer tends to vary with the standpoint of the commentator. While IMF experts stress macroeconomic, fiscal and monetary policies, bilateral aid negotiators stress administrative competence and industrial economists stress technical competence on the part of the state apparatus.[27] But what they all agree on is that there is a conflict between Kenyan capital that was 'entrenched' in the early post-independence years, especially in commerce (by Kenyanization policies and by ISI itself), and the needs of the manufacturing sector now.[28] In particular, people with political influence illegally import goods that undercut the market of a vulnerable local manufacturer, while other manufacturers, who have monopolies, use their political influence to resist measures intended to liberalize imports and make them more efficient.[29] The senior staff of large, inefficient

[25] Diana Hunt, *The Impending Crisis in Kenya: The Case for Land Reform* (Aldershot: Gower, 1984), p. 288. Hunt's assumptions can of course be challenged, but no other equally large-scale means of absorbing surplus labour has yet been proposed.

[26] Godfrey, *Kenya to 1990*, op. cit., p. 79. Kenyan wages were reduced in real terms by about 20% between 1981 and 1985.

[27] Coughlin, 'Toward a New Industrialisation Strategy', op. cit., p. 9 and *passim*.

[28] For a pioneering study of 'entrenchment' in Africa, see Gunilla Andrae and Björn Beckman, *The Wheat Trap* (London: Zed Press, 1985).

[29] See, for example, *Africa Research Bulletin*, 30 November 1987.

parastatals (who include some large independent accumulators, especially through landownership) resist their privatization or the introduction of more competition. Politicians resist fiscal discipline in order to channel resources to their constituencies or into the pockets of themselves and their backers. For example, one of the main contributors to the budget deficit of 1986, which totally broke the government's commitments to the World Bank and the IMF, was £33 million spent on the All Africa Games, of which the auditor said £14 million were lost through inefficiency and corruption.[30]

In good years international capital is apt to feel sympathetic toward the Kenya government's political dilemmas, and especially its sensitivity to public opposition to measures that cut real incomes or aggravate inequalities, and it has been known even to smile indulgently at populist extravagances, such as Moi's £10 million bill in 1988 to celebrate ten years of his own rule. But at other times it expresses its anxiety in stern language: 'The question remains whether the country's leaders have the political will to push through the necessary reforms while Kenya has the resources to undertake them [i.e. the population–land crisis has not yet come to a head], or whether they will be swayed by short-term gains.'[31]

On the whole, the aid community has got what it wanted in Kenya, and the other side of the coin is that, for its part, it has also been willing to help stave off political threats to further capital accumulation by providing enough external finance to allow the economy to go on expanding. The consultative group has consistently met most of the Kenya government's requests, making it one of the largest per capita recipients of external capital flows in sub-Saharan Africa and one of the first debtor countries to benefit from the debt cancellations following the 1988 OECD summit in Toronto (US$435 million of debt was cancelled by West Germany, about US$200 million by France, US$178 million by the USA, US$150 million by the UK, US$109 million by Canada, US$20 million by the Netherlands, and about US$10 million by Belgium — even though Kenya did not fall in the category of most indebted countries that the Toronto initiative was intended to help). And, on the other side of the coin again, the Moi government's consistently pro-Western foreign policy was an obvious counterpart to this.

The main contradiction of capitalist development, then, as the aid community sees it, is as follows: As the population grows, there is a growing risk of catastrophic increases in deprivation, leading to political instability and the disappearance of the conditions for a rate of capitalist growth capable of outpacing the growth of needs. The option of land reform, which would buy time and widen the domestic market for manufactures, is blocked by the interests of the large landowners who are essential to the capitalist development strategy. The resulting necessity to rely all the more heavily on manufacturing exports confronts numerous secondary contradictions. It means overcoming the entrenched interests of various powerful segments

[30] *Africa Research Bulletin*, 15 November 1988.

[31] *Africa Research Bulletin* (citing *The Times* and the *Financial Times*), 30 November 1988.

of the domestic capitalist class, and it means imposing further cuts in the real incomes of wage earners. But failure to move in this direction carries the risk that capital accumulation will falter. This is already a threat. Through the 1980s private foreign investment stagnated (a significant number of foreign firms actually withdrew — 25 out of 140 US companies, in fact), and there was a great deal of covert flight by domestic capital, to the point where the aid community expressed concern that Kenyans' transfers of funds abroad were significantly offsetting the very substantial inflows of aid. In 1987 the IMF estimated that US$1.2 billion was held by Kenyans in non-bank accounts overseas; the Kenyan government responded by prosecuting a few Asian businessmen for exchange violations.[32]

It is clear that the government's reluctance to impose further costs on the rest of society for the sake of manufacturing — whether further reductions in real wages and cuts in services for the poor or the denial of profitable import licences and the like to its wealthy backers — was not unconnected with the fact that Africans (as opposed to Kenyan Asians) were still virtually unrepresented in manufacturing. Coughlin's conclusion, based on research into several sectors of manufacturing in the mid-1980s, is hard to dispute:

> Kenyan manufacturing industry is almost exclusively owned by MNCs, Kenyan Asians, or government parastatals; Africans own very few medium or large-sized manufacturing firms. This has seriously impeded any identification of interests between local industrialists and the political circles. As a result, the government's economic policies and bureaucratic decisions are frequently detrimental to the nation's long-term industrialisation. Just occasionally do politicians oppose and defeat colleagues who arrange lucrative but improper deals with importers... Only after more Africans own and manage manufacturing enterprises and ensure that their economic interests are well represented politically will policy become better harmonised with the needs of industry.[33]

On the other hand, it was also important to retain and reassure the Kenyan Asians, whose capital and managerial and technical skills were such an important reason for the relative efficiency of the non-agricultural sectors of the Kenyan economy. When in 1986 President Moi called for more indigenization, there was no sign that it was because he had grasped the strategic importance of identifying African capital with manufacturing. It seemed rather to pander to racial resentment, and it aggravated Asian insecurity.[34]

[32] *Africa Research Bulletin*, 31 October 1987, 30 November 1987. Coughlin, 'Toward a New Industrialisation Strategy', op cit., gives various examples of the way funds are transferred abroad, in spite of measures adopted in the 1960s and 1970s to prevent this.

[33] Coughlin, 'Toward a New Industrialisation Strategy', op cit., p. 16.

[34] 'The talk of "Kenyanisation" has now turned to "indigenisation"... Kenyans interpret the word "indigenisation" as excluding local whites and Asians. Yet the 70,000-strong Asian community, already jittery about the remote possibility of a *coup d'état*, dominates important sectors of the Kenyan economy' (*Africa Research Bulletin*, 31 January 1987).

This general contradiction — i.e. between the needs of capitalist development and the political conditions for it, including the entrenched interests of the Kenyan capitalist class — appears to be inherent in the process of capitalist development itself, at least in the late twentieth century and in a country such as Kenya. There are, however, other aspects of Kenyan policy in the 1980s which are less easily accounted for. Two in particular stand out: Kenya's relations with its neighbours, Uganda and Tanzania, and the use of political power to acquire assets or to punish and impoverish critics, often at the cost of considerable economic losses.

With regard to the first, the collapse of the East African Community in 1977 cut Kenya's manufactured exports to Tanzania from over K£20 million in 1975 to a mere K£1800 in 1978.[35] The reopening of the border between the two countries in 1983 led to a strong recovery of exports to Tanzania by 1985, but in the meantime the losses had been costly, and the cause is generally seen as having been mainly ideological — i.e. dislike of Nyerere's 'African socialism' coupled with an unrealistic expectation that Tanzania would agree to remain in a common market in which Kenya ran a large and permanent trade surplus with both its neighbours. A similar lack of sensitivity seems to have marked Kenya's relations with Uganda following the accession to power of Museveni's National Resistance Movement in 1986, with the harassment of Ugandans in Kenya and periodic semi-blockades of Uganda's imports through Mombasa, pushing Uganda to search seriously for an alternative route through Tanzania. In both cases a dysfunctional political animus, compounded by national chauvinism, seems to have governed policy-making, rather than a concern for the good trading relations that would maximize the prospects for Kenyan manufactured exports over the long run.[36]

With regard to the use of political power to acquire assets or punish critics, this is for obvious reasons less fully documented, even when it is well known, although some famous cases do get reported. In a well-publicized case in 1989, the main maize-milling company, Unga, stopped production for several weeks because it was prevented from getting essential supplies in the course of a struggle to force its African owners to cede a share of control to others closer to the president. Many similar cases, often ending in bankruptcies and unemployment, could be cited.[37]

[35] *National Development Plan, 1989 to 1993* (Nairobi: Government Printer, 1989), p. 7.

[36] Ugandans believed that the 1987–88 hold-up of Ugandan imports was a means by which some powerful Kenyans who had invested in trucks to transport Ugandan coffee exports sought to thwart the Uganda government's policy of diverting these exports back to the much cheaper railway (*Africa Research Bulletin*, 31 January 1988).

[37] One case that symbolizes them all occurred in Kisii in the mid-1980s. Two women's groups had obtained power-driven grain mills through the assistance of foreign non-government organizations (NGOs). One group had been helped to secure this assistance by a former MP and critic of the regime, George Anyona. At the ceremony to inaugurate the installation of the mill, the police came, broke up the concrete foundation and removed the mill to the Kisii police station to make it clear that such benefits were not to be had except through patrons approved by the regime.

What these practices show is that the 'predatory capitalism' that marked the 1960s and early 1970s under Kenyatta was strongly renewed under Moi as the Kalenjin notables sought to enrich themselves by means similar to those used by the Kikuyu notables under Kenyatta. Whether or not there had been a growing desire among the Kikuyu capitalist class to transcend this kind of behaviour, with the chronic insecurity it generated, in the final years of the Kenyatta regime, its resumption by Moi gave it a new lease of life. It set back the development of a more productivist bourgeois culture. For example, far from upgrading the skills and autonomy of the judiciary (a matter of increasing concern to leading Kenyan capitalists in the mid-1970s), in 1988 a constitutional amendment removed the security of tenure of High Court and Appeal Court judges and made them liable to dismissal by the president acting alone. Moreover, these changes came 'at a time when some members of Kenya's judiciary [had] voiced fear about increasing political influence over legal decisions'.[38] What is more, the new 'predatory capitalism' was pursued at the expense of Kikuyu capitalists, more than of non-Africans, as had been the case in the 1960s and 1970s, aggravating ethnic divisions between Kenyan capitalists.

This draws attention to what is perhaps the most fundamental weakness of the Kenyan African capitalist class, looked at from the angle adopted here. The evidence indicates clearly that it was far from having 'formed itself into a class', far from having organized itself around a shared project for the transition to sustainable capitalist development in Kenya. On the contrary, it remained deeply permeated and divided by ethnic consciousness, still strongly reliant on political influence (rather than market competitiveness) for its profits and security, and still rather ignorant of and unsympathetic to the needs of modern industry, hardly any of which it owned. It tolerated the non-African Kenyan capitalists but was not inclined to make any sacrifices for their interests. The productivist values of the precolonial Kikuyu accumulators brought to light by Cowen may have been important, but they had been diluted, to say the least, and somewhat overlaid by other values and practices since colonial times. Rather than developing political institutions capable of organizing its collective class interests and attaching other classes to itself through political leadership, it continued to rely on Bonapartism, ceding political power to a single individual in return for having all other possible centres of power subordinated to him.[39] And the Bonaparte in question was no strategist of capitalist development.

As a result, at the end of the 1980s no prospect of a solution to the fundamental or underlying problem of capitalist development in Kenya was in sight or even being discussed, and many of the secondary contradictions discussed above remained equally unresolved. What Godfrey had

[38] *African Research Bulletin*, 15 September 1988.

[39] See Peter Anyang' Nyong'o, 'State and Society in Kenya: The Disintegration of the Nationalist Coalitions and the Rise of Presidential Authoritarianism, 1963–1978', *African Affairs* 88, 1989, pp. 229–51.

written in 1985 could have been repeated almost word for word in 1990: 'In general, the prospect is of scraping through quite successfully for another five years even though the fundamental problem facing the economy [the rising labour/land ratio] remains largely unaddressed.'[40]

Yet, relative to other African countries, Kenyan development remained buoyant, and the 1980s business literature on Kenya reflected this ambivalence, commenting on the 'context of relative optimism', on the 'good' prospects for manufacturers, noting some of the surprisingly large advances that have been achieved in particular areas (e.g., a 25% growth of manufactured exports in 1988), judging that social inequalities will not lead to political instability in the medium run, and so on. Partly this reflects the relatively short-run perspective of much international capital. It also partly reflected a general conviction that Kenya would not be 'allowed', in the immediately foreseeable future, to go into a destabilizing economic crisis, given its importance for the strategic interests of the USA. This belief was supported by the facts. So long as the Moi regime's repression of all opposition seemed compatible with further accumulation, little criticism of it was voiced by the aid community states, in spite of the official post-Helsinki emphasis on human rights in Communist regimes. But, once Moi's repression began to get out of hand, with the murder of his foreign minister (who was suspected of being about to disclose the extensive corruption of many of his colleagues), the disaffection of the intellectual and professional middle class, and growing popular demonstrations calling for democracy, the aid community applied firm pressure on Moi to make concessions, culminating in his abrupt decision in late 1991 to accept multiparty elections for early 1992.[41]

The literature also, I suspect, reflects a general assumption that, other things being equal, there is a dynamic tendency in the capitalist system for problems to be overcome — in one way or another. But the literature's concern with political stability registers the fact that the political threat to continued accumulation is permanent and unpredictable. It has been well said that, in economics, one is never looking over the edge of the abyss: the market always levels the future off to a gradual slope. No such benign mechanism is at work in politics. The 'impending crisis' of which Diana Hunt wrote, the 'underlying problem' of the aid community in relation to Kenya, is one where the social tensions caused by deepening landlessness and unemployment can be contained, in the longer run, only by diverting resources indispensable for sustaining growth.

None the less, the outcome is not predetermined, and in the meantime

[40] Godfrey, *Kenya to 1990*, op. cit., p. 90.

[41] Donor pressure was first concerted at the consultative group meeting in December 1990 and reached its peak in August–September 1991, when both Norway and Denmark had halted their aid programmes and others were threatening to (*Africa Research Bulletin*, November–December 1990 and September–October 1991). The British government naturally (in view of British capital's special place in Kenya) dragged its feet and continued publicly to support the Moi regime's monopoly of power as late as September 1991.

what needs to be done is to study the social forces that are being brought into being by capitalism in Kenya — and not least the Kenyan capitalist class, whose conflicts and collaborations will determine how the problem is eventually articulated, in what forms it presents itself, what responses are offered, and what they imply for the accumulation process. The Kenyan bourgeoisie, seen through the unsentimental eyes of international capital, does not appear in a particularly flattering light; but it does not appear uniformly incompetent either, or incapable of learning, and it does not have to shoulder the task of resolving the contradictions of capital accumulation in Kenya unaided — it has interested and willing foreign collaborators. It is not written in stone that capital will not find solutions to the problems that confront it in Kenya — at whatever cost to those who do not own any.

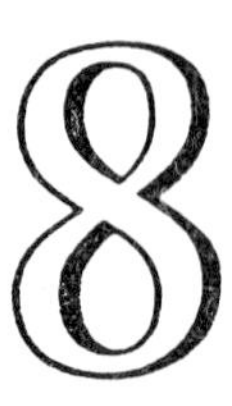

African Capitalists & Development

However development is defined, it must involve the accumulation of capital. Only out of the surplus saved from past productive effort can any society obtain a larger sum of the values for which development is pursued — more health or education, more leisure or more output.[1] In any developing society, therefore, someone must ensure that production is organized so as to yield the necessary surplus, and see also that this surplus is set aside from immediate consumption and invested appropriately. In capitalist societies this is done by capitalists and their managers, supported, more or less directly, by the state. The modern multinational corporation (MNC) is clearly the most highly refined and potent agency of capital accumulation the world has ever seen but, even in their 'home' countries, such as the USA or Japan, the tasks of accumulation are not performed by them alone; and in most underdeveloped countries MNCs typically operate only in a limited number of sectors, such as soft drinks, beer and oil distribution, and perhaps construction or mining. So, for capitalist development to occur, there must also be local, domestic, internal, 'national' (and, perhaps, 'indigenous') capitalists; and these vary greatly in their individual and collective capacities — the scale of the capital they dispose of, the technical and organizational skills they command, the social cohesiveness they exhibit, the political power they wield, the ideological influence they enjoy, and so on. In the absence of any practicable alternative to capitalist development, therefore, it becomes very important to understand what determines the relative strengths and weaknesses of each underdeveloped country's internal or domestic capitalist class.

[1] More equality can, in theory at least, be obtained from redistribution without any increase in capital or material output, as the example of Cuba shows; and this can produce considerable increases in both individual and collective values such as education, health, etc. There is no obvious reason why this should not be seen as a kind of development, but hitherto redistribution alone has not been thought of as development, which accounts for the definition adopted in the text.

Moreover, as soon as one looks at capitalists in terms of their overall role in development, it is their political as much as their economic and technical capacity that seems important; in other words, it is as social classes, acting more or less consciously in their collective interest *vis-à-vis* other social classes, that they enter decisively into history and critically affect the prospects for their countries' development. As social classes, however, African capitalists have not been extensively studied.

There are several possible reasons for this. In the first place, fully formed African capitalist classes have been the exception rather than the rule. The formation of a politically influential and productive capitalist class, with a solid and mature bourgeois culture to support and focus its economic and political projects, has never been the work of a generation or two. Even the relatively industrialized countries of the former Soviet Union are now discovering this to their chagrin — to acquire a class of parasitic *nouveaux riches* is one thing, but a class of serious wealth creators is quite another; and in most parts of Africa the domestic capitalist class is still, at best, embryonic and weak.

Second, mainstream political science in the USA, which until recently tended to dominate African political science, has consistently rejected the idea that social classes should be central objects of study.[2] It was not until the end of the 1960s, when the intellectual initiative in development studies passed to dependency theory, that the existence and importance of social classes and class politics in Africa began to be more widely acknowledged. But this happened at a price: the work that began to be done on African capitalist classes in the 1970s was seriously limited — indeed deformed — by its origins in the debate about dependency. The chief question asked was whether African capitalist classes really existed and, if so, how independent they were from foreign capital, as opposed to being compradors — mere conduits or agents of foreign capital, with no capacity or will to enforce their own independent interests at the expense of the interests of foreign capital or to promote 'national' capitalist development.

This question was largely misconceived, although the studies it stimulated and the controversies surrounding them have added a good deal to our knowledge. The question was misconceived in that it begged a prior question. It assumed that the chief cause of Africa's underdevelopment, including the weakness of its internal capitalist classes, was the power of foreign capital, whereas it would be truer to say that the weakness of African capital was simply part and parcel of Africa's underdevelopment and that foreign capital filled (however inadequately and with various attendant costs) the vacuum created by the absence of strong domestic capitalist

[2] For an interesting account of the ideological bias of US development studies in the 1950s and 1960s, see Irene Gendzier, *Managing Political Change: Social Scientists and the Third World* (Boulder and London: Westview Press, 1985).

classes in Africa.[3] The capacity of the local capitalist class varies considerably from country to country in Africa: and it is far from obvious that, where its capacity is low, this is because foreign capital is particularly strong. We need instead to enquire what are the important characteristics of an internal capitalist class for the process of capital accumulation — what are the key capacities it needs for national development to go forward — and then to look at the historical conditions favouring or not favouring the acquisition of these capacities by a local class of accumulators.

But in order to do this we need to clarify our theoretical terms. There has been an understandable reaction against general theory, after the exciting but inconclusive debates about modernization and dependency in the 1970s, and in face of the seemingly relentless economic decline and political decay in sub-Saharan Africa since that time. But, since 'data' are never really 'given' but are always defined by theoretical presuppositions, we must still try to be clear about the inherited conceptual tools we are using. The first part of this chapter, therefore, briefly reviews the main theoretical and conceptual reference points for any study of capitalist classes in Africa today.

Marx and Weber on capitalist classes

It is natural to suppose that Marx, who made social class the hinge concept of his theory of history — the concept linking economic development to political action — provided a framework for analysing capitalist classes that we can apply to Africa; but in reality only limited help is to be found in his work. This is mainly because, in his most systematic treatment of the topic, Marx took an almost purely structuralist view of the bourgeoisie, declaring, in the preface to the first edition of *Das Kapital*, that according to his standpoint individuals were of interest only in so far as they were 'the personifications of economic categories', the creatures of social relations, not their authors.[4] He took England as the *locus classicus* of the capitalist mode of production up to that time and said that less developed countries could expect to follow the same path in due course, as the 'natural laws of capitalist production' (which he admittedly often also called only tendencies) 'won their way through', 'worked themselves out with iron necessity'.

In this famous passage he seems to leave no room at all for the performance of a particular country's capitalist class to make any difference. Of course, elsewhere he frequently alluded to the difference that the course of class struggles could make, and in his political writing (especially on France in the late 1840s and early 1850s, but also on Britain and elsewhere) he analysed the different fractions of capital, the interplay of economic and

[3] This point seems only too consistent with the fact that widespread disinvestment in Africa by MNCs in the 1980s did not lead to a notable strengthening of local capitalist classes or to an acceleration of national development.

[4] Karl Marx, *Capital* (London: Penguin Books, 1976), Vol. 1, p. 92.

political power, the complexities of capitalist ideology, and so on. But it is hard to deny that the dominant theme in his thinking is the unfolding logic of capital accumulation and above all its impact on the workers: the prime determinant of the outcome of the class struggle is 'the development of the working class itself', not that of the bourgeoisie. That a capitalist class had emerged and become economically (if not politically) dominant in Britain was obvious; that capitalist classes would become similarly dominant in Germany and France was something he assumed.[5] If a capitalist class did not achieve power, because of the resistance of precapitalist classes, this would presumably mean 'the common ruin of the contending classes' (as he and Engels put it in the *Communist Manifesto*).

In Marx's account there is, then, little allowance for differences of degree or quality between the various national capitalist classes making a difference to how far they would become dominant and with what effects. Marx could be scathing about the political limitations of this or that national capitalist class,[6] but he never tackled the general question of what makes some capitalist classes more effective than others, or what makes a given capitalist class succeed at one time and fail at another. As is well known, Engels near the end of his life confronted the general question of the role of agency in history in his letters to Schmidt and Bloch, but with notoriously debatable results. And, in spite of the extensive debate that has taken place about the limitations of the British capitalist class, few of the participants have taken up the issue systematically or comparatively; they have not asked what are the general theoretical implications of acknowledging the extent to which the political and technical incapacity of the manufacturing fraction of the British capitalist class has hampered the later capitalist development in Britain.[7]

Yet two aspects of Marx's concept of the capitalist class deserve to be emphasized and retained: first, its clear place in the general theory of the

[5] Marx originally seems to have expected the capitalist class to become dominant everywhere. But in the late 1870s he had some second thoughts, provoked by the Russian populists who wondered if Russia really had to go through capitalism to get to socialism; and in 1881, after much reflection, he said that his earlier view that capitalism was historically inevitable was restricted to the countries of Western Europe. In Russia (and by implication perhaps elsewhere) it might be prevented and socialism might be attained on the basis of indigenous peasant institutions, if these institutions could be protected from the 'harmful influences assailing [them] from all sides' (see Marx's famous letter to Vera Zasulich of 8 March 1881 and the discarded earlier drafts for it, in Teodor Shanin's highly relevant study, *Late Marx and the Russian Road: Marx and 'the Peripheries of Capitalism'* (New York: Monthly Review Press, 1983)).

[6] As Perry Anderson has pointed out, whereas in the *Communist Manifesto* Marx wrote that the 'modern bourgeoisie' (namely, the British) had conquered 'exclusive political sway', in later life he began to see the British bourgeoisie as 'self-limiting and subordinate in its [political] actions and aspirations' ('The Figures of Descent', *New Left Review* 161, January–February 1987, p. 24).

[7] Eric Hobsbawm, in his *Industry and Empire* (London: Penguin Books, 1969), followed Marx's structuralist approach as far as possible (and beyond), viewing British capitalists as following the 'logic of capital' but in conditions inherited from an [continued on page 168]

capitalist mode of production, so that a capitalist class is a class of people who occupy similar places in capitalist relations of production, i.e. appropriating surplus value and then having the possibility and the necessity to invest this surplus in ever more productive ways, to generate further surplus value; and, second, the need of this class to 'conquer' state power, to have the state serve its needs in reorganizing and regulating society. Taken together, these features make the bourgeoisie a class that makes history, as both the product of the logic of capitalist accumulation and its agent. Capitalists are driven to do what they do by the logic of the market ('accumulate, accumulate, that is Moses and the prophets'), but they also need to act collectively to defend and promote their interests against those of other classes. These two things, the one producing constant transformations in production (productivity-raising changes), and the other producing periodic transformations in social relations through state power (the bourgeois revolution that clears the way for the productivity increases to be sustained), make the capitalist class significant.

These points are worth stressing because of the serious dilution that the concept of class often suffers in the literature on Africa. Any category of people may be called a class, but, unless the concept belongs to a theory with some potential explanatory power, labelling them in this way adds nothing of value. Marx did not provide a general theory of the ways in which capitalist classes perform their historic function, but he indicated the sort of theory it needed to be.

As for Weber, his thought has deeply marked US scholarship on Africa, both generally, through its influence in US social science, and specifically, through the influence on the 'modernization' school of Talcott Parsons, Weber's translator and interpreter. But, as Thomas Callaghy has shown, Weber offered a much richer and more sophisticated set of ideas for understanding the development of capitalism and the role of capitalist classes than is to be found in Parsons's mechanical and ahistorical apparatus of 'pattern variables'.[8] It is not possible to recapitulate these ideas, which Callaghy succinctly outlines, in the space available here. The central point is that Weber's ideal type of 'modern capitalism' is one in which

earlier period that led, in spite of their profit-maximizing behaviour, to a general loss of international competitiveness. Harold Perkin and Martin Wiener, adopting a Weberian approach, saw British capitalists as having abandoned their original 'entrepreneurial ideal' or 'industrial spirit' in favour of the antimaterialist and antiscientific outlook of the old precapitalist political class — i.e. they did not see them as forced by the 'logic of capitalism' to behave in any ideal way (see H. Perkin, *The Origins of Modern English Society 1870–1880* (London: Routledge, 1969) and M. Wiener, *English Culture and the Decline of the Industrial Spirit* (Cambridge: Cambridge University Press, 1981)). Neither side in this debate offered a general theory of the role of capitalist classes in development.

[8] Thomas M. Callaghy, 'The State and the Development of Capitalism in Africa: Theoretical, Historical and Comparative Reflections', in Donald Rothchild and Naomi Chazan (eds), *The Precarious Balance: State and Society in Africa* (Boulder and London: Westview Press, 1988), pp. 67–99. Although Callaghy's own Weberian approach differs from the one adopted here, his argument shows significant parallels.

'instrumental rationality' is applied to every aspect of production and the state; and this further entails that many other spheres of life must also be subjected to the same rationality, or at least be made compatible with it. So, not only must markets, the organization of work, accounting, taxation and public administration exhibit this rationality, but also science, justice, morals, religious belief and much else. Consequently, a key area of enquiry must be the extent to which the capitalist class is able to make these numerous spheres of life exhibit or at least not resist instrumental rationality; and this in turn raises important questions about their political and cultural influence, the extent to which they are both dominant and hegemonic, i.e. enjoy both power and consent. And this in turn involves above all the capitalist class's relation to and influence over the state, which is, generally speaking, the only agency capable of reorganizing or at least regulating all these spheres of life.

Unlike Marx, however, Weber does not propose a causal theory. He declares a wide range of social, economic and political factors to be important for the rise of capitalism, but does not suggest which are most important or propose that they become important in any particular sequence. His view is 'non-deterministic, non-evolutionary'; what happens is the outcome of struggles and hence indeterminate.[9] This is considered by many scholars (including Callaghy) to be a merit on Weber's part, although Weber too seems to have imagined that, in the long run, capitalism would become universal.

Marx, of course, thought that the logic of capitalist competition acted as a 'revolutionizing' hidden hand, forcing capitalists into certain courses of action, both individually as capitalists trying to survive in the market and collectively as a class trying to advance its interests. This relentless ongoing pressure, he thought, tended to result in the installation of the capitalist mode of production and the dominance and hegemony of the capitalist class. And, speaking generally, it is hard not to agree that this has been the tendency worldwide. On the other hand, whether or not any given country experiences this global tendency as 'beneficial' or 'progressive' — in the historical sense intended by Marx himself (meaning liberating people from drudgery, advancing them to new levels of productivity, breaking their parochial chains, giving them access to the world's cultural and scientific heritage and so on) — is another matter. Capitalism does not develop in every region similarly. There are poles of growth and margins of decay; areas of 'high mass consumption' and other areas that are mere pools of surplus labour, vegetating on the periphery or even sinking into social disintegration, famine and anarchy. So what is needed is a theory of the role of the capitalist class in determining which kind of outcome occurs in particular kinds of case.

It is tempting, but certainly premature, to imagine that out of the work that is now being done the elements of such a theory will emerge, in a

[9] Callaghy, 'The State and the Development of Capitalism in Africa', op. cit., p. 73.

judicious blend of the ideas of Marx and Weber. For the moment, all that seems possible is to keep such a theory in view as a horizon — a sort of moving target that one knows one will not reach, but that serves to orientate and inspire empirical research that may in due course contribute to building the theory we need.

Development theory

It is to the credit of the modernization school, with its strong Weberian roots, that questions about domestic or internal capitalists were fairly central to it from the outset. The weakness of this approach is that it rejected the concept of social class in favour of that of an 'élite' and hence did not raise the issue of whether or how the internal capitalist class might be the means of transforming the social relations of production. Instead, the issue posed was how far there were internal candidates for the élite roles proper to a capitalist economy and polity, as though these roles merely had to be occupied for capitalist development to forge ahead. The general theme of the modernization literature was that business élites were essential to development and that factors characteristic of traditional society, especially 'traditional values', inhibited the emergence of such élites. This was a central theme of the work of Everett Hagen and Bert Hoselitz, to name two of the most influential modernization theorists; and David McLelland, with Weber's thesis on the role of the Protestant ethic in mind, even developed the idea that the level of 'achievement orientation' in any culture could be scientifically measured, and that development could be accelerated by 'injecting' businessmen in developing countries with some of the 'achievement factor' that was missing in their cultures.[10] Fred Riggs and Daniel Lerner explored the ways in which societies in transition from precapitalist to capitalist relations of production offered resistance to the application or absorption of capitalist rationality by both the class of capital and the state, in terms that strikingly anticipated the World Bank's rediscovery of the same phenomenon in Africa in the late 1980s.[11]

[10] Everett Hagen, *On the Theory of Social Change: How Economic Growth Begins* (Homewood, Illinois: Dorsey Press, 1962); Bert Hoselitz, *Sociological Aspects of Economic Growth* (New York: Free Press, 1960); David McLelland, *The Achieving Society* (Princeton: Van Nostrand, 1961). The US government sponsored training programmes based on McLelland's ideas for Third World businessmen.

[11] Fred Riggs, *Administration in Developing Countries: The Theory of Prismatic Society* (Boston: Houghton-Mifflin, 1964) and *Prismatic Society Reconsidered* (Morristown, New Jersey: General Learning Press, 1973); Daniel Lerner, *The Passing of Traditional Society: Modernizing the Middle East* (New York: Free Press, 1958). It is noteworthy that in the late 1980s disappointed modernization theorists produced metaphors very similar to Riggs's prism to describe the so-called neo-patrimonial state in Africa, diverted from instrumental rationality by the 'economy of affection' and displaying simultaneously the features of the modern, technically-efficient 'air conditioner' at the front and the traditional, personalistic 'verandah' (where clients' needs are catered for) at the back. See Goran Hyden, *No Shortcuts to Progress: African*

Although no useful theory seems to have emerged from this literature, at least it paid attention to some real issues affecting the formation and efficacy of domestic, internal or indigenous capitalist classes. In contrast, dependency theorists, in their attack on the modernization school, seemed at first relentlessly structuralist. The capitalist classes of the periphery were pictured as mere comprador agents of metropolitan capital, thanks to the global structures of dependency of which they were 'bearers'. This was particularly true of the influential early writing of Andre Gunder Frank;[12] although, as Frank was later to point out (and as the subtitle of his first book indicated), his work as a whole was almost entirely based on the work of Latin American historians and was not particularly structuralist, except in its first chapter (which, however, was what made the greatest impression, with its grand schema of 'three contradictions' and the rest). In his subsequent writing, Frank insisted that the 'lumpenbourgeoisies' (as he called them) of Latin America were the product of conflicts in which embryonic parties or tendencies pushing for a less dependent, more nationalist development strategy had been defeated, with the help of outside (imperialist) interests.[13] In this version, the domestic capitalist classes of Latin America figure as more active agents of underdevelopment, but their 'lumpen' (i.e. unprogressive, comprador) character still appears as a more or less unavoidable result of the structures of dependency, which also predetermine that this fraction of the domestic capitalist class will prevail.

Less fatalistic versions of dependency theory, such as that of Fernando Cardoso and Enzo Faletto, or the 'bureaucratic authoritarianism' thesis of Guillermo O'Donnell, saw the character and role of the internal capitalist class as much more determined by the specific historical circumstances of each country.[14] In particular, these later products of the dependency school focused on critical economic transitions, such as the transition from primary commodity exports to industrial production, or from import-substitution industrialization to 'deepening' the industrial sector, seeing these as involving radical realignments both within the domestic capitalist class and

Development in Management Perspective (Berkeley and Los Angeles: University of California Press, 1983); and Emmanuel Terray, 'Le Climatiseur et la véranda', in *Afrique plurielle, Afrique actuelle: Hommage à Georges Balandier* (Paris: Karthala, 1986), p. 38, cited in Donal B. Cruise O'Brien, 'The Show of State in a Neo-colonial Twilight: Francophone Africa', in James Manor (ed.), *Rethinking Third World Politics* (London: Longman, 1991), p. 151. Scientific progress seems to have been modest in this sector. The World Bank's later thinking may be found in *Sub-Saharan Africa: From Crisis to Sustainable Growth: A Long-Term Perspective Study* (Washington: World Bank, 1989).

[12] A.G. Frank, *Capitalism and Underdevelopment in Latin America: Historical Studies of Chile and Brazil* (New York: Monthly Review Press, 1967).

[13] A.G. Frank, *Lumpenbourgeoisie, Lumpendevelopment: Dependence, Class and Politics in Latin America* (New York: Monthly Review Press, 1972).

[14] Fernando Henrique Cardoso and Enzo Faletto, *Dependency and Development in Latin America* (New York: Monthly Review Press, 1967); Guillermo O'Donnell, *Modernization and Bureaucratic Authoritarianism: Studies in South American Politics* (Berkeley: Institute of International Studies, University of California, 1979).

in its relations with other domestic classes and with foreign capital and the military. Indeed, these versions of dependency theory clearly combine themes from both Marx and Weber in the way previously suggested, and are quite suggestive for the study of African capitalists — not so much in detail (since the historical contexts of Latin America and Africa are so different) as in the kind of theorizing they attempt.

Although Marxist critics of dependency theory, such as Geoffrey Kay and Bill Warren, attacked its structuralism, they did not put in its place any theory of the determinants of the course of peripheral development in general, or the varying roles of indigenous capitalist classes. Their tendency is always to look to Marx's analysis of the logic of capital accumulation for a sufficient explanation of the behaviour of capitalists, both as individuals and as a class. This does not prevent them from offering some valuable insights into the question. For instance, Kay's theory of the shift from the dominance of foreign commercial capital to that of foreign manufacturing capital as an explanation of underdevelopment in the African colonies is an interesting set of hypotheses of a type that, like those of O'Donnell, might usefully be brought to bear on the evolution of African capitalists; and it has some echoes in the contemporary literature on rent-seeking in Africa, and also in Gunilla Andrae and Björn Beckman's thesis about the vesting of particular capitalist interests as a barrier to further expansion based on 'market-rationality', all of which seek to find rational explanations for types of capitalist behaviour that tend to block further capital accumulation for the economy as a whole.[15]

The comparative history of capitalism

Until recently, scholars interested in nascent capitalism in the underdeveloped world have usually looked for parallels (or contrasts) in the history of capitalism in its area of birth, i.e. Western Europe, where David Landes and other historians of European industrialization have commented on the varying capacities, technical and political, of the various European bourgeoisies. Landes is particularly relevant for the study of African capitalist development because he tries to extract from comparative history the necessary and sufficient conditions for industrial capitalism to become

[15] See G.B. Kay, *Development and Underdevelopment: A Marxist Analysis* (London: Macmillan, 1975), Chapter 5; and Bill Warren, *Imperialism: Pioneer of Capitalism* (London: Verso, 1980). Kay's argument was that manufacturing capital was gradually able to squeeze the profits of commercial capital, as a supplier of inputs to manufacturing from the colonies; and, because for various reasons commercial capital was not able to reorganize production in the colonies so as to increase the amount of surplus value created there, it could maintain its own profit only by exerting ultimately disastrous downward pressure on the prices it paid for the African peasants' produce. On rent-seeking, see Mark Gallagher, *Rent-seeking and Economic Growth in Africa* (Boulder: Westview Press, 1991). On vesting, see Gunilla Andrae and Björn Beckman, *The Wheat Trap* (London: Zed Books, 1985).

established anywhere. He lays a good deal of stress on the high value that was placed in Europe, by the eighteenth century, on the 'rational manipulation of the human and material environment', as indicated by such things as the scope for rationality afforded by Christianity (i.e. even by Catholicism) relative to other religions such as Islam. And in passing he endorses in general terms Weber's hypothesis about the effect of the Protestant ethic on the rise of capitalistic behaviour. He also stresses the widespread interest in science in Europe; and, for Europe in general and Britain in particular, he attaches particular importance to the way in which entrepreneurial values and practices were able to become established in various kinds of social 'space' that was insulated, in one way or another, from the surrounding society. Thus, the city could be a space shielded from the surrounding rural society, with its traditional order and values; or religious discrimination could force minorities into business careers by excluding them from public life (for example, Dissenters and Jews in Britain).[16] This is a theme of some relevance in contemporary Africa, as we shall see. In contrast, Alexander Gerschenkron focused on the agencies and, above all, the states that played such a key role in the industrialization of backward economies in the nineteenth century (i.e. backward in relation to Britain's) — a theme also much debated today.[17]

But it is striking how little these and many other historians have tried to tie together these two aspects: the culture and special circumstances that give rise to an industrializing bourgeoisie, and what determines whether state power either complements or frustrates capitalist industrialization. In Carlo Cipolla's *Economic History of Europe*, for example, Claude Fohlen comments on the limitations of the entrepreneurial outlook of a large section of the French capitalist class in the 1870s, and Knut Borchardt comments on the monopolistic character of late eighteenth-century German industrialists, dependent for their success 'too often on procuring courtly extravagance or the compulsory recruitment of labour'.[18] But the treatment of the characteristics of these national capitalist classes is quite unsystematic in these texts and no interest is expressed in their influence on the policies pursued by their governments. Important changes in state economic policy appear as 'exogenous' changes to the environment in which these capitalists operated, to which they themselves are presented as having made no contribution at all. Of course, in the early stages of capitalist development, it can frequently happen that state economic policy reflects the interests

[16] David Landes, *The Unbound Prometheus: Technological Change and Industrial Development in Western Europe from 1750 to the Present* (Cambridge: Cambridge University Press, 1969), especially pp. 21–23 and 66–74.

[17] See Alexander Gerschenkron, *Economic Backwardness in Historical Perspective: A Book of Essays* (Cambridge, Mass.: Harvard University Press, 1962).

[18] Knut Borchardt, 'The Industrial Revolution in Germany 1700–1914', in Carlo M. Cipolla (ed.), *The Fontana Economic History of Europe: The Emergence of Industrial Societies* (Glasgow: Fontana-Collins, 1973), Vol. I, p. 87; see also Claude Fohlen, 'The Industrial Revolution in France 1700–1914', in ibid., pp. 7–75.

and wishes of the capitalist class only indirectly, if at all; but from the first they naturally seek to influence policy, and the state can seldom ignore them altogether. Yet little or nothing is said in these texts about the ways in which different fractions of capital — financial, commercial, industrial or agricultural, for example — conceived of their collective interests and organized to defend them, or tried to ally themselves with other social classes to get a 'regime of accumulation' in which they could prosper.

The case is different, however, with the much more recent literature on capitalism outside Europe and especially the work that has been stimulated by the dramatic rise of the newly industrializing countries (NICs) in East Asia. In these cases, the role of the domestic capitalist class is absolutely central. Korea, especially, has been the focus of intense interest and the formation of its industrial capitalist class is particularly interesting and instructive, involving as it does a complex mix of deep, long-term cultural and political traditions; the distinctive and radical socio-economic transformations wrought by Japanese colonialism; the special role of the post-war US military government and US aid; and the forced transformation of the rent-seeking commercial bourgeoisie into a manufacturing class by the state under the Park regime after 1961.[19]

Merely to list these moments in the formation of the Korean capitalist class is, of course, to show how removed the Korean experience has been from that of sub-Saharan Africa. And, in spite of its 'newness', Korean industrialization was already far advanced in the 1930s, when in most of Africa it had not even begun. But even the sharp contrasts are instructive, particularly the South East Asian countries' centuries-long history of unified governments, shared languages and religions, powerful landowning classes, etc. Above all, the story that is now beginning to be told about these countries foregrounds their capitalist classes — in their evolving interaction with the state, other classes, foreign technology, corporate organization, cultural development and the rest — in a way that marks it off from the story hitherto told about most European countries.[20] In this literature there is no mystery about the class forces behind, or in opposition to, state economic policy; and future research in Africa should be able to benefit considerably from it.

[19] See especially Carter J. Eckert, *Offspring of Empire: The Koch'ang Kims and the Colonial Origins of Korean Nationalism* (Seattle and London: University of Washington Press, 1991), for an exemplary analysis of the formation of Korea's industrial capitalist class under Japanese rule; Clive Hamilton, *Capitalist Industrialization in Korea* (Boulder and London: Westview Press, 1986), especially pp. 29–50; also Alice H. Amsden, *Asia's Next Giant: South Korea and Late Industrialization* (New York: Oxford University Press, 1989).

[20] No doubt this does an injustice to many writers who have studied the development of capitalism outside its European and North American homelands well before the recent fascination with the NICs. An interesting example of such earlier work is Norman Jacobs, *The Origin of Modern Capitalism and Eastern Asia* (Hong Kong: Hong Kong University Press and Oxford University Press, 1958), which seeks to explain, by means of a detailed comparison of Japan and China, why 'not all societies accept capitalism, even when that capitalism "knocks at the door", as Weber and others allege' (p. 213).

Research on capitalist classes in Africa

With the exception of a few early studies by Marxists — mainly attempts to characterize the whole spectrum of social classes in particular countries, such as Samir Amin's work on Senegal, Mahmoud Hussein's work on Egypt or Mahmoud Mamdani's study of Uganda — little work on capitalist classes as such (i.e. as opposed to studies of businessmen or business élites) seems to have been undertaken before the advent of the debate on dependency in Africa.[21] Some material on indigenous capitalists was, of course, also generated by the debates — precursors of the dependency debate — about neo-colonialism in Ghana and Tanzania, debates partly stimulated and strongly influenced by Frantz Fanon's reflections in *The Wretched of the Earth* (London: MacGibbon and Kee, 1965) (reflections that were partly based on what he saw happening in Ghana when he was in Accra in 1960 as ambassador for the provisional government of Algeria).[22]

But in the 1970s, when the question was posed in a more general way by the debate over dependency, research began to be undertaken that (to a greater or lesser extent) focused primarily on domestic capitalist classes *qua* classes — i.e. not merely on the quality of the entrepreneurship of the individuals who composed them, but also on the extent to which they had developed a consciousness of their collective interests and the organizational means to promote them politically. Leading examples of such studies are Rhoda Howard's historical analysis of the Ghanaian bourgeoisie, Fatima Mahmoud's study of the Sudanese bourgeoisie, Malaak Zaalouk's study of the Egyptian bourgeoisie, Nicola Swainson's study of Kenyan capitalists[23] and Michael Schatzberg's study of the bourgeoisie in rural Zaïre.

[21] Samir Amin, *Le Monde des affaires sénégalais* (Paris: Editions du Minuit, 1969); Mahmoud Hussein (a pseudonym), *La Lutte des classes en Egypte* (Paris: Maspero, 1968); Mahmoud Mamdani, *Politics and Class Formation in Uganda* (New York: Monthly Review Press, 1976).

[22] Examples are Bob Fitch and Mary Oppenheimer, *Ghana: End of an Illusion* (New York: Monthly Review Press, 1966); Roger Genoud, *Nationalism and Economic Development in Ghana* (New York: Praeger, 1969); and Issa Shivji, *Class Struggles in Tanzania* (London: Heinemann, 1975). All these authors argued that, contrary to their own self-representation, these regimes were capitalist but, given the limitations of the domestic capitalist class (or the real absence of such a class), were incapable of achieving 'independent' national economic development.

[23] Rhoda Howard, *Colonialism and Underdevelopment in Ghana* (New York: Africana Publishing, 1978); Fatima B. Mahmoud, *The Sudanese Bourgeoisie: Vanguard of Development?* (London: Zed Books, 1984); Malak Zaalouk, *Power, Class and Foreign Capital in Egypt: The Rise of the New Bourgeoisie* (London: Zed Books, 1989); Nicola Swainson, *The Development of Corporate Capitalism in Kenya 1918–1977* (Berkeley: University of California Press; London: James Currey, 1980). Swainson's study was not exclusively concerned with the Kenyan African capitalists, but was conceived in the context of the dependency debate and was the basis for her strong antidependency stance in her article 'The Rise of a National Bourgeoisie in Kenya', *Review of African Political Economy*, 8, January–April 1977, pp. 39–55. The same issue of the *Review* carried an important early article on the same theme by Paul Kennedy, 'Indigenous Capitalism in Ghana' (pp. 21–38). Côte d'Ivoire and Kenya, the focus of much of the debate, served as a prime focus for the conference that resulted in Paul Lubeck (ed.), *The African Bourgeoisie: Capitalist Development in Nigeria, Kenya and the Ivory Coast* (Boulder: [continued on page 176]

Two collections of the late 1980s highlighted this whole period of work: Irving Leonard Markovitz's *Studies in Power and Class in Africa* and Paul Lubeck's *The African Bourgeoisie*.[24]

The main thrust of all these studies was to establish how far the local or domestic capitalist class was 'independent' of foreign capital, with the explicit or (more often) implicit assumption that such independence was the key to successful national capitalist development. The result is an indispensable body of information, but the costs of their shared preoccupation with dependency theory — whether the evidence was seen as confirming or refuting it — were considerable.

The pro-dependency studies tended to focus on the relative limitations of Africa's capitalist classes and to ignore their growth paths over time, especially their collective ability to learn (both technical and managerial skills, and political skills). These studies were apt to understate the accomplishments of the African capitalist classes and to set (implicitly) impossibly high standards for them, relative to the length of time it actually took the capitalist classes of today's industrialized countries to become effective social forces. They tended (like the Russian populists a hundred years earlier) to see capitalism as something occurring only in the visible, formal sector of the economy and to overlook the ongoing gradual but crucial transformation of rural relations of production into increasingly commercial and finally capitalist relations (with a market for land, and wage-labour employed in commodity production).[25] They also tended to overlook the corresponding expansion in urban areas of non-enumerated or informal economic activities of an increasingly capitalist nature. (It is interesting to reflect how little of the process of capitalist development in Britain, capitalism's original heartland, would have been captured by any contemporary study in, say, the early eighteenth century that had focused only on what is today considered the formal or enumerated sector.) And there was a tendency to view the political role of the domestic capitalist class purely in terms of whether it 'controlled' state power — and, if so, whether it used this power to reduce the dominance of foreign capital in the economy — rather than to ask how far its politics were conducive to overcoming the real

Lynne Rienner, 1987). Valuable collections of citations may also be found in Henry Bernstein and Bonnie K. Campbell (eds), *Contradictions of Accumulation in Africa* (Beverly Hills: Sage Publications, 1985), especially in the chapters by Björn Beckman, Gavin Kitching and Bonnie Campbell. See also Jean-François Médard, 'The Historical Trajectories of the Ivorian and Kenyan States', in Manor (ed.), *Rethinking Third World Politics*, op. cit., pp. 185–212.

[24] I.L. Markovitz was among the first to grasp the growth and significance of African capitalist classes in his 1977 book, *Power and Class in Africa* (Englewood Cliffs, New Jersey: Prentice-Hall, 1977), Chapter 7. His edited volume of 1987, *Studies in Power and Class in Africa* (New York: Oxford University Press, 1987), followed up on this work; Parts 2 and 3 are particularly relevant.

[25] On the very parallel debate between the Russian populists and Lenin, see the excellent overview by Gavin Kitching, *Development and Underdevelopment in Historical Perspective* (London: Methuen, 1982), Chapters 2 and 3.

obstacles to sustained capital accumulation in the country, with or without the assistance of foreign capital.

Conversely, the opponents of dependency theory tended to assume that, if a local capitalist class could be shown to exist (and, better still, if it exercised a significant measure of political power *vis-à-vis* foreign capital), then the future of capitalist development was more or less assured — or at least could be presumed not to be hopeless.[26] They mostly did not critically examine the bases of accumulation of the domestic capitalist class (for instance, the extent to which it depended on politically acquired privileges or monopolies, rather than its own entrepreneurial, financial and technical competence in the market-place); the determinants of its internal cohesion and capacity for collective political action (for instance, how far it transcended ethnic boundaries); or its possession or non-possession of a shared concern for establishing the general conditions needed for further accumulation to be possible (i.e. how far it had a class 'project'). They tended to oppose to the *dependencistas*' pessimism about the prospects for capitalist development an equally ill-founded optimism.

After the dependency debate: contemporary research

With the advantage of hindsight, especially with the disastrous evolution of Africa's economies since the late 1970s (when most of the information in the above-mentioned studies was collected), it is possible to pose the issues in a more open way. Several illusions that tended to haunt the debate about dependency have been destroyed, such as the dream of an alternative socialist development path waiting to be taken (if only a 'genuinely radical' African leadership would take it), or the fantasy of a smooth and socially benign alternative path of 'genuinely independent' or 'progressive' national capitalist development — dreams or fantasies against which researchers often implicitly contrasted the corrupt, socially dislocating, ugly and immiserating experience of 'actually existing' capitalist development.[27] All of this can now be replaced by a serious question: in any given country, what are the obstacles confronting a sustainable process of development of the forces of production under capitalist relations of production and how do the historically given characteristics of the country's internal capitalist

[26] As Beckman pointed out, this was roughly the case with the article I wrote that initiated the 'Kenya debate': see Colin Leys, 'Capital Accumulation, Class Formation and Dependency: The Significance of the Kenyan Case', *Socialist Register*, 1978, pp. 241–266; and Björn Beckman, 'Imperialism and Capitalist Transformation: Critique of the Kenya Debate', *Review of African Political Economy* 19, 1980, pp. 48–62. The debate is reviewed in Chapter 7 above.

[27] The expression 'actually existing capitalism' is intended to echo the expression 'actually existing socialism' with which Rudolph Bahro sought to get the European left to confront the reality of the neo-Stalinist systems of Eastern Europe in the 1970s (Rudolph Bahro, *The Alternative in Eastern Europe* (London: New Left Books, 1978)).

class bear on the possibilities of overcoming these obstacles?

This is not to say that questioning the moral and political implications of capitalist development is pointless, or that alternative paths of development are not worth seeking. On the contrary, as I will argue in the conclusion to this chapter, such questions seem more urgent than ever. But, in order to raise such questions, the scope and limits of actually existing capitalist development in Africa have first to be properly understood and for this we have to understand the past, present and potential capacity of Africa's capitalist classes.

Two general books published in the 1980s do express the general point of view that has just been outlined: John Iliffe's *The Emergence of African Capitalism* and Paul Kennedy's *African Capitalism: The Struggle for Ascendancy*.[28] Both authors essentially relegate the dependency debate to the sidelines and take it as given that capitalism is developing in Africa, as elsewhere. For them, the interesting questions concern what factors affect the speed and character of capitalist accumulation, either in Africa generally or in particular regions or countries in Africa and with what results.

Iliffe sees the distinguishing characteristics of Africa's experience of capitalism as arising from the combination of the distinctive features of the continent's precapitalist societies with the 'very late stage in the global history of capitalism' that had been reached when capitalism penetrated Africa.[29] Indigenous African capitalism, which had existed here and there before colonialism, was overwhelmed by competition from advanced capital in the metropoles, backed up by colonial rule. Indigenous capitalism re-emerged under colonialism, in agriculture (primarily through a gradual process of differentiation among smallholders engaged in export commodity production), in trade and finally in industry. Iliffe offers a fascinating overview of the way religions, both indigenous and imported, have both helped and hindered capital accumulation by Africans; and in conclusion he explores the way African capitalists have fared at the hands of independent African states. This experience he divides into three broad types: anti-capitalist regimes (such as Nkrumah's Ghana or Nyerere's Tanzania); 'parasitic capitalism', as in Zaïre, where state power is used primarily to take capital from those who have organized its production; and 'nurture capitalism', in which the state works, ostensibly at least, to support capital accumulation by Africans (as examples he offers Kenya and Côte d'Ivoire).

[28] John Iliffe, *The Emergence of African Capitalism* (London: Macmillan, 1983); Paul Kennedy, *African Capitalism: The Struggle for Ascendancy* (Cambridge: Cambridge University Press, 1988). Another book of the same vintage and with similar scope is John Sender and Sheila Smith, *The Development of Capitalism in Africa* (London: Methuen, 1986), but it is powered by an antidependency perspective that leads to external obstacles to capitalist development being minimized and to no distinctions being drawn between the various African countries in terms of the prospects of enjoying capitalist development. Consistently, there is no discussion of the distinctions to be made between the continent's various capitalist classes.

[29] Iliffe, *The Emergence of African Capitalism*, op. cit., p. 4.

Iliffe is a historian. He concentrates on the origins of capitalist accumulation by Africans, without trying to predict success or failure for their efforts. Kennedy, on the other hand, is a social scientist with a historical approach. In his view, the dependency debate about the relative strengths of domestic and foreign capital was ahistorical; he believes that closer study of at least some African capitalist classes suggests that they are at a stage of development comparable to that of their European precursors not twenty, but two hundred, years ago.[30] He then notes the many respects in which the conditions for rapid accumulation in Africa are less promising than they seem to have been in England (for example) two hundred years ago.[31] But he sees no reason to suppose that the incentives of wealth and status, combined with foreign capital's need for stable regimes underpinned by local collaborators, will not make it possible for African capitalists to keep moving up an entrepreneurial and political learning curve.[32]

Of course, Kennedy's book was written before the full scale of the recession and the so-called global restructuring of capitalism had become apparent and his views may now be judged overoptimistic. But, as an antidote to unfounded pessimism and a reminder of the persistent pressure and opportunity to make a profit that gives capitalism its distinctive dynamism — and its capacity for creativity as well as for destruction — it at least helps redirect attention back to the wider questions about the capacities of African capitalist classes, which both sides in the immediately preceding literature tended to ignore.

Among Kennedy's numerous suggestions that seem to invite further research, two seem especially worth emphasizing. One is that rising bourgeois classes have always had to make alliances and compromises with other classes and to share state power with them. Rarely have capitalists actually held 'exclusive political sway' (the rhetoric of Marx and Engels notwithstanding) and least of all when they were still at an early stage of development as a class. This means that it is almost as important to understand the extent to which other classes or social forces have congruent interests with those of the capitalists, or can be induced to support the policies advocated by the local bourgeoisie, as to understand the characteristics of the local bourgeoisie itself.[33]

Another suggestion to be pondered concerns the importance that anthropologists and others have attached to 'outsider' status of various kinds in the emergence of capitalist accumulation.[34] Membership in a 'traditional'

[30] Kennedy, *African Capitalism*, op. cit., pp. 112–34.

[31] Ibid., pp. 99–102, 147–56.

[32] Ibid., pp. 184–91.

[33] Ibid., pp. 87–88. The concept of sharing state power is no doubt almost as unsatisfactory as that of controlling it. For one of the few authoritative applications of modern state theory to Africa see Bruce J. Berman, *Control and Crisis in Colonial Kenya* (London: James Currey, 1990).

[34] The role of certain groups, some of which were 'outsiders', was also a celebrated theme of Werner Sombart's writings on capitalism, especially the role of national traits — which, however, he attributed to entrepreneurial genes in the blood of the Frisians, the Jews, and the lowland Scots (*The Quintessence of Capitalism* (New York: Dutton, 1915), pp. 200–21).

group can help an individual capitalist to accumulate, but as often as not it is not being a member of the locally dominant traditional group (linguistic, religious, cultural, etc.) that provides the freedom from group pressure necessary for accumulation based on the exploitation of labour. That is, we are looking at African equivalents to the kinds of 'space' seen by Landes (see p. 173) as being so important for some of the early shoots of capitalism to take root in Europe.[35] Past a certain point, Kennedy notes, the ambient social structure eventually reasserts limits to the accumulation process on the part of 'outsiders' — which is another way of saying that the conditions that permit the beginnings of accumulation may not permit its further development but become fetters on it, unless and until they are themselves changed (traditional power gives way to new forces).

In this context, another shortcoming of the literature on African capitalism prompted by the dependency debate can be more easily identified and overcome: the uncritical identification of domestic or indigenous capitalism with 'African' capitalism. This is a severely misleading elision. In Africa, as in Europe and elsewhere, it has often been domestic and strictly indigenous groups that are in one sense or another outsiders (in the African cases, 'racially' defined outsiders) who have been responsible for much of the early development of capitalism — and whose separate status may then itself become a problem for continued accumulation in the post-colonial period.

So far, rather few studies of African capitalist classes have been undertaken from the point of view represented by Kennedy's work and advocated here. A notable exception, however, is Janet MacGaffey's 1987 study of the capitalists of Kisangani, Zaïre, *Entrepreneurs and Parasites*.[36] Although her fieldwork was done in the late 1970s, her perspective transcended the preoccupations of the times and offered an open-minded and detailed account of the development of a domestic bourgeoisie, in the full sense of the term — i.e. not confined to the sphere of circulation (trade) and not dependent on political patronage and state monopolies, but rather emerging — in spite of the destructive greed of the 'parasites' in positions of state power — in the economic space opened up by the very destruction of the state's capacity to administer and regulate the economy that the parasites' corruption has brought about.

MacGaffey did not discuss the prospects for replacing the dominance of the parasites with the hegemony of the relatively productivist entrepreneurial class represented by the businessmen and businesswomen of Kisangani; nor did she consider whether any project for national development was

[35] Ibid., pp. 137–43.

[36] Janet MacGaffey, *Entrepreneurs and Parasites: The Struggle for Indigenous Capitalism in Zaire* (Cambridge: Cambridge University Press, 1987). Another valuable monograph on class development in rural Zaïre is Michael G. Schatzberg, *Politics and Class in Zaïre: Bureaucracy, Business and Beer in Lisala* (New York: Africana Publishing, 1990). Schatzberg's focus, however, is on what he terms the 'national politico-commercial bourgeoisie', the parasites of MacGaffey's title, which could hardly be considered a bourgeoisie at all.

likely to emerge from this quarter. On the face of it, it did not appear likely. The capitalist class she described seemed to suffer from the contradiction that success on a significant scale would re-attract the fatal attention of the parasites and lead to another round of plunder, bringing accumulation once more to a standstill. On the other hand, her account reminds us of the resilience of the domestic capitalist classes in Africa and the dynamic potential of the accumulation process to restart itself, once a certain commercialization of the economy has occurred. It also suggests that, in changed political circumstances, such as the regime change that was still being pursued in Zaïre — however faint-heartedly — by the 'donor community' in 1992–93, a new class of actually existing capitalists might acquire a measure of political influence and help set in motion a reversal of the long decay of the Zaïrean economy.

And, at the very least, MacGaffey's work draws a long-overdue line between real capitalists and the politicians, civil servants and colonels who use their political power to get rich — Fanon's 'bourgeoisie of the state'. For many years, observers have imagined that the use — too often the abuse — of office in order to accumulate assets must be regarded as a route, however unsavoury, by which African capitalists will in due course emerge. Parallels have been drawn between the plunder of gold from India and the Americas in the late eighteenth and early nineteenth centuries and the looting of the public treasuries in post-colonial Africa, as merely different kinds of 'primitive capital accumulation'. But, while some individual capitalists may have emerged, or may yet emerge, from the families enriched by state corruption in Africa, in the medium run this possibility seems of trivial importance compared with the spectacular damage this same corruption has wrought, both to the infrastructure on which production depends and directly on the production process itself, including the production of many genuine capitalists who have been stripped of their capital by these same 'kleptocrats'. As Callaghy has aptly remarked, 'Why call it [the so-called organizational or bureaucratic bourgeoisie] a bourgeoisie at all? The continued use of the term simply leads to conceptual confusion.'[37]

Of course, even if we were to accept that in Zaïre the vitality of its provincial domestic capitalist class holds out a possibility, however slight, of future capitalist development, we cannot generalize from Zaïre, a very large and resource-rich economy by African standards. Many other African countries would not necessarily hold out much promise of accelerated capitalist development even if they proved to have internal capitalist classes with the capacities of those found by MacGaffey in Kisangani. And, even in the case of Zaïre, the question still has to be asked: if the internal capitalist class remains unable to prevent the state from sabotaging capitalist development

[37] Callaghy, 'The State and the Development of Capitalism in Africa', op. cit., p. 85. Callaghy prefers the term 'political aristocracy', by analogy with prerevolutionary France. This perhaps reflects an American's ingrained republican distaste for aristocrats. I prefer the term 'state kleptocracy' as being less complimentary.

(let alone make it actually support it), can the wider conditions for capitalist development identified by Weber ever really be secured? What MacGaffey's work does show, however, is the importance of looking objectively at the nature of the domestic classes that have emerged as one very important term in the development equation in each individual African country and region.

An agenda for future research

One way to sum up what has been argued here is to distinguish between two broad functions that need to be performed by capitalists: the organization of production and exchange, and the organization of the political conditions for the continued accumulation of capital. Under the former we can also distinguish between the direct organization of production and exchange by individual capitals (not excluding co-operation between them, as in cartels, etc.) and the organization of infrastructural services necessary for production, usually (although not necessarily) provided by the state.

In theory, the direct organization of production could be undertaken entirely by foreign capital, but in practice much of what can be produced locally is unlikely to be profitable for foreign capital. Likewise, the provision of infrastructures need not be done by the state, but only very large-scale infrastructural projects or services are likely to be profitable for foreign capital, and others tend to prove difficult areas for efficient co-operation by smaller-scale, local capitals. So the role of the domestic capitalist class is important in relation to these functions. It makes a big difference whether it performs them, or ensures their performance by the state, well or badly.

Similarly with the political function. Circumstances may occasionally exist when external forces can assume responsibility for securing the political conditions for continued capital accumulation; and the deep involvement of aid donors in economic policy-making in Africa in the era of structural adjustment programmes draws them inevitably into politics, to the point where cases of near-recolonization are increasingly numerous.[38] But, in the long run, this is a function for which a domestic capitalist class is normally crucial.

But the domestic capitalist class does not necessarily perform either its productive or its political functions well, or even at all. What needs to be investigated is how far it performs them and what conditions determine this. The particular historical circumstances affecting its emergence and

[38] The intervention of donors (who are in this context mostly not donors but lenders concerned about the security of their loans) in Kenya in 1991–93 is a classic example. Having obliged the government to hold elections, which it won, the donors then obliged it to accept a package of monetary reforms by withholding aid for a further four months. The World Bank's vice-president for Africa commented: 'We have managed to reach agreement and the government has come all the way to the bank' ('Penitence Wins Kenya Return of World Bank Aid', *Guardian*, 22 April 1993).

formation into a class need to be established, to discover the determinants of its technical, organizational and financial competence and the scope of its activities in the direct organization of production and exchange; its technical and organizational competence with respect to the construction and expansion of the infrastructure; and its political solidarity and skills (in coalition building, establishing ideological hegemony and neutralizing opposition).

This implies quite an extensive agenda for future research:

1. Identifying the different components of the domestic capitalist class — by economic sectors, or 'fractions', or other categories into which it is divided — and examining their respective historical origins and growth paths.

2. Studying the way these origins have influenced the fields of accumulation they have entered and their 'business cultures', including their technical and organizational capacities, and assessing these capacities in relation to the requirements of further accumulation. How far does the class depend for continued accumulation on state-awarded privileges (monopolies, contracts, quotas, licences, etc.) alone?

3. Examining the internal economic, social, political and ideological relations of the class. How far does it depend on the internal relations within particular social sub-groups (ethnic, male, etc.) for its economic security and prospects of further accumulation? Or, to put it another way, how impersonal, instrumental-rational have the class's internal relationships become? And how far has it begun to develop a collective class consciousness that transcends its component parts and how far does this consciousness include a class political project or vision of the future for society as a whole? What institutions — political, social, cultural, etc. — manifest the stage reached in the formation of the class and its distinctive character?

4. Studying the relations between the internal capitalist class and other classes, especially the petty bourgeoisie, old and 'new.'[39] How far are they, or might they become, relations of bourgeois hegemony? How far has the domestic bourgeoisie acted in such a way as to induce other classes to accept, as in their interests, the requirements for continued capital accumulation? How far has such a conception become, or might it become, influential in determining state policy?

5. Studying the relations between the domestic capitalist class, within the system of class relations as a whole and other systems of authority, power and status that intersect with it, including gender relations.

6. Examining the relations between the domestic bourgeoisie and the state

[39] This usage is not very satisfactory, but was effectively popularized by Nicos Poulantzas in his *Political Power and the Social Classes* (London: New Left Books, 1973). By the 'old' petty bourgeoisie is meant the class of producers who own their own means of production but who use it themselves and do not rely exclusively on waged employees to produce the output — i.e. peasant farmers, artisans, traders. By the 'new' petty bourgeoisie is meant the class of educated employees, mainly in state employment (especially in Africa), who have something in common ideologically with the old petty bourgeoisie.

apparatus, with regard to the economic conditions for capital accumulation (infrastructure, fiscal policy, trade policy, management of foreign exchange and debt, etc.) and the political conditions for it (treatment of allied interests, regulation of labour, land policy, etc.). How developed, rational and institutionalized are these relations (as opposed to personalized, subjective and vulnerable to change)? How far is the state, actually or potentially, capable of instituting the necessary social and political preconditions for sustained capital accumulation?

7. Studying the relations between the internal bourgeoisie and foreign capital and the agencies of foreign capital (bilateral and multilateral funding agencies, in particular). How far are these relations collaborative, rational and mutually advantageous? How far has the domestic capitalist class become recognized by these external actors as a significant and reliable partner in the development process?

8. Studying decisive moments and events in the evolution of the internal capitalist class that have affected its subsequent economic and political capacity — i.e. the course of its class struggles. As in individual biographies, the study of 'turning-points' can powerfully illuminate all the important dimensions of its subsequent trajectory.

9. Studying the obstacles to further capitalist development. What are the main obstacles, on what do the prospects for overcoming them mainly depend, and how far is there a shared perception and understanding of these issues within the domestic capitalist class?

The contemporary context

A final consideration concerns the medium-run global context in which African capitalists must operate today. Every country, and so every domestic capitalist class, confronts not only a unique set of domestic problems, but also a partly unique external context — a specific historically given set of trading, financial, military, diplomatic and other relationships, within which any future development must occur. An oil-exporting country such as Gabon, a former Frontline state in the Southern African liberation struggle such as Zambia, a country dependent on a single low value-added export crop such as Senegal — each has a different external environment on account of the distinctive features of its geography and history. Yet one aspect of the external environment is broadly common to all African countries, and in conclusion it seems important to touch briefly on this because, contrary to the received wisdom of only a decade or two ago, it makes the task confronting their internal capitalist classes so much harder. I refer to the growing productivity gap between agricultural production in all still predominantly precapitalist agrarian economies and capitalist production elsewhere.

The point here is a simple one, but it rarely seems to get the sober attention it deserves, perhaps because it is so troubling. All African countries without exception depend on being able to sell primary commodities for export. These exports must compete in international markets with the produce of other countries, in a growing number of which capitalist relations of production prevail or are at least well established in certain sectors. This means that the value of these commodities is constantly falling, in the sense that less and less labour time is required for their production where they are produced under capitalist conditions of production. Unless demand rises, therefore, the price falls; and in countries where the labour time expended falls more slowly, or even stays constant (not to mention rises, which can happen if the infrastructure or other circumstances affecting production deteriorate), this means that the return to this labour falls too. Eventually, the return to labour can fall below the level at which people will produce the commodity at all.

So stated, the point may seem abstract and theoretical. But something like this has actually been happening to one African agricultural commodity after another: palm oil, groundnuts, cocoa, coffee, cotton and many other crops are now produced many times more productively by large-scale mechanized agriculture on large farms, estates or plantations in Malaysia, the USA, Brazil, Egypt, Sudan and elsewhere, to the point at which export production by African smallholders is increasingly under threat, if not already finished (as in the case of Nigerian palm oil, for example).[40] And what capitalist farmers cannot produce more cheaply, capitalist manufacturers will naturally try to produce cheaper substitutes for and will aggressively market these as replacements (e.g. synthetic textiles).

Historically, of course, at the same time as capitalism has raised agricultural productivity and by the same token reduced agricultural employment, it has opened up alternative employment in industry. However, it has seldom done both things in exactly the same place or at the same pace; the process has mostly involved very painful social dislocations, eased by large-scale migrations, often international (e.g. from Europe to North America) and over several generations.

The problem confronting Africa today is that, on the one hand, this kind of large-scale migration overseas is ruled out, while, on the other, industrial productivity is itself advancing so rapidly that the capacity of manufacturing to create alternative employment has drastically declined. Whereas it used to be imagined that African countries, as late industrializers, would

[40] For a further development of this point see Chapter 6 above. Of course, supply and demand do not stay constant and here the problem is aggravated by the fact that all underdeveloped countries are being urged simultaneously by the World Bank to expand their exports of primary commodities, leading to still weaker prices. The effect is that governments find themselves trying to boost output by offering guaranteed producer prices that can only be maintained by subsidies they have no other adequate source of revenue to pay for. For the case of cocoa in Ghana, see Cord Jakobeit, 'Reviving Cocoa: Policies and Perspectives on Structural Adjustment in Ghana's Key Agricultural Sector', in Donald S. Rothchild (ed.), *Ghana: The Political Economy of Recovery* (Boulder: Lynne Rienner, 1991), pp. 221–32.

be able to profit from the hard-won productivity gains of their predecessors embodied in the productive technology now available to them, the reverse now seems more true; it would be much easier for Africa to industrialize if average industrial productivity worldwide had remained at the level it was at 150 years ago. As a result, African populations are becoming increasingly marginal to global production.

Marx analysed the 'relative surplus population', the production of which he quite rightly saw as capitalism's most distinctive effect (what he called the 'general law of capitalist accumulation'), into four categories: (i) a 'floating' element, which we would perhaps today call frictional unemployment, i.e. people temporarily out of work due to shifts in supply and demand, changes in technology, new products and so on; (ii) 'latent', by which he meant agricultural producers whose destiny was sooner or later to be forced off the land due to rising agricultural productivity; (iii) 'stagnant' — the unskilled but able-bodied urban unemployed, ex-peasants not absorbed into productive non-agrarian labour; and (iv) 'pauperism' — the permanently unemployed and largely unemployable urban pool of orphans, disabled, sick, etc., plus the lumpenproletariat — vagabonds, criminals, prostitutes, etc.[41] Using these categories, we may say that in Africa the first category is barely significant, the second category is unprecedentedly vast and the condition of the third category is already so desperate that in many countries it increasingly shades off into the fourth and last category. (And Africa has, of course, produced some new forms of pauperism to add to Marx's list, such as the continent's 3.5 million refugees, the tens of thousands with missing limbs or eyes in Mozambique and Angola, the Somalis and Mozambicans and others permanently dependent on food centres operated by various foreign agencies, the villages in Uganda and other central African countries devastated by AIDS, and the marauding gangs of armed teenagers who threaten to become a permanent feature of life in large areas of Mozambique.)

What all this means is that the political problem confronting any African capitalist class, even if it can find solutions to the complex technical and economic problems involved in expanding capitalist production, often seems close to being insoluble. And yet, with a further twist of the screw, there is pressure, as much from donors as from Africa's populations, for democratization, which was introduced in most of today's industrial countries only after the violent transition from precapitalist agrarian society to capitalist industrial society had been largely completed and accepted and when wage employment had begun to succeed in absorbing the displaced peasantry.

These sombre reflections are not intended to imply that capitalist development is nowhere possible in Africa. Paul Kennedy's insistence on capitalism's dynamic potential needs to be kept in mind. But in studying the capacity of Africa's capitalist classes to play their distinctive role in African

[41] Karl Marx, *Capital* (London: Penguin Books, 1976), Vol. 1, pp. 794–98.

development, we must face realistically the difficulties they confront in the shape of world capitalism as it actually exists and not as it existed fifty or a hundred years ago. In some African countries they may overcome the obstacles and succeed in making them into productive capitalist economies. In others, marginalization and social distress may lead either to recolonization (via successive forms of aid programmes, including military aid, for example) or to various kinds of anticapitalist reaction — fundamentalist or nativist, perhaps, or some new forms of socialism — whose contours cannot as yet even be imagined. The only thing that seems reasonably certain is that understanding Africa's capitalist classes will remain important for understanding whatever happens next.

9

Development Theory & the African Tragedy

It is at least fifteen years since the literature began to be full of references to an African 'crisis'. Today it seems more appropriate to use another word, not just because 'crisis' has been so overworked in so many contexts, but because Africa is not in fact balanced on a knife's edge between recovery or collapse: it is a tragedy that is already far advanced. Millions of people have already died from hunger, disease and violence, and millions more face Hobbesian existences in conditions of accelerating environmental and social degradation: famines, chronic malnutrition, the collapse of health services, the erosion of education, reappearing endemic and epidemic diseases, AIDS, endemic criminal violence, civil wars, genocide — the capitalism-induced barbarism of which Rosa Luxemburg warned.[1] These are facts of the African tragedy: the issue is not whether they will happen, but whether they can be prevented from getting worse, and gradually brought to an end.

Little optimism is to be found in the existing literature: very few people believe that things in Africa are going to get better. The World Bank, which in the past was consistently overoptimistic about African economic prospects, acknowledged in its 1989 long-term perspective study of sub-Saharan Africa that the 'nightmare scenario' that had been sketched by the United Nations (UN) Economic Commission for Africa in 1983 was 'a very real possibility'.[2] Assuming that fertility rates fall over the coming years as they have in South Asia (a somewhat optimistic assumption), the World Bank expects that the population of sub-Saharan Africa will none the less double from about 500 million in 1990 to 1 thousand million in 2010, and reach 1.7 thousand million by 2050.[3] The pressure portended by these figures on

[1] 'The destruction of culture...the dictatorship of militarism...dissolution and an early decline into chaos and anarchy...a string of political and social disasters and convulsions...punctuated by periodical economic catastrophes or crises.' See Norman Geras, *The Legacy of Rosa Luxemburg* (London: New Left Books, 1976), pp. 21–35.

[2] World Bank, *Sub-Saharan Africa: From Crisis to Sustainable Growth, a Long-Term Perspective Study* (Washington, DC: World Bank, 1989), p. 33.

[3] Ibid., pp. 32–33. The Bank's population projections have confirmed this estimate even when the devastation of deaths from AIDS are taken into account.

already scarce fertile land and on already crumbling urban infrastructures can only be dimly imagined. To provide these people with productive employment and to prevent competition for ever-scarcer resources from leading to increasingly unmanageable political conflict, the region's economies need to grow, the World Bank observed, by 'at least 4 to 5 percent a year', i.e. at least twice the 2.1% annual growth achieved in the region in 1985–88, not to mention the 1.2% annual decline experienced in the years 1981–84.[4]

The prospects for achieving this target, however, do not seem good. The actual annual growth rate for the region in the years since 1989 (according to World Bank figures for 1990–92), was less than 3% (i.e. per capita incomes were still falling).[5] These years may prove unrepresentative: things may improve. But it is not obvious that the requirements specified by the World Bank for reaching a growth rate of 4–5% are more likely to be met in the future (the Bank itself regarded them as 'ambitious'). They included much higher levels of investment (25% of the region's GDP, a 66% increase over the 1988 level), based on higher levels of domestic saving and drastically increased capital flows from outside Africa; and much better use of capital (by the 1980s the average rate of return to investment in Africa had fallen to about 10% of its 1960s level).[6]

Official development assistance to sub-Saharan Africa did nearly meet the World Bank's *Berg Report* (*Accelerated Development in Sub-Saharan Africa* (Washington: World Bank, 1981)) target of US$17.8 billion by 1990, reaching US$16.5 billion in 1991 (although it levelled off at US$16.2 billion in 1992, and was of course offset by annual debt service payments by sub-Saharan Africa of roughly the same amounts). On the other hand, the 1989 study also called for private capital to account for most of the investment required in African agriculture and industry, to the tune of 7% of the region's total GDP, a target which seemed increasingly unrealistic by the mid-1990s.

With regard to the rate of return on investment, 'structural adjustment' should in theory improve the efficiency of capital use, but as structural adjustment programmes have lived up to few of the virtues claimed for them it is hard to be optimistic on this score.[7] Conditionality may be reducing the gross misuse of resources by ministries and parastatals, but the continuing deterioration of the infrastructure (which structural adjustment

[4] Ibid., pp. 4 and 33.

[5] Israt Husain, 'Structural Adjustment and the Long-term Development of Sub-Saharan Africa', in R. Van der Hoeven and Fred Van der Kraaij (eds), *Structural Adjustment and Beyond in Sub-Saharan Africa* (The Hague/London: Ministry of Foreign Affairs/James Currey, 1994), p. 157, citing World Bank internal data.

[6] World Bank, *Sub-Saharan Africa*, op. cit., p. 26.

[7] See, for example, Paul Mosley and John Weeks, 'Has Recovery Begun? "Africa's Adjustment in the 1980s" Revisited', *World Development* 21/10, 1993, pp. 1583–606; and Thomas M. Callaghy and John Ravenhill (eds), *Hemmed In: Responses to Africa's Economic Decline* (New York: Columbia University Press, 1993).

policies have tended to aggravate) tends to counteract any gains from this change by making all investments less productive. For all these reasons, few observers seem really to believe that a recovery is in prospect; yet if the decline persists the catastrophes of hunger, dislocation and violence that Africans have already experienced are liable to become more frequent and more widespread.[8]

As for the explanations of the tragedy that have been put forward, what strikes one in retrospect is that there is some truth, at least, in every one of them. It is undeniable that 'neo-patrimonialism', 'rent-seeking' and the exploitation of ethnic differences have commonly characterized African politics, leading to a systematic misuse of public resources. It is true that African presidents and their associates have often appropriated public funds, sometimes on a massive scale, siphoning off the surpluses needed for investment (or even for paying public service salaries) and, in the process, legitimating extortion and corruption throughout the system.[9] It is also beyond question that, even in the absence of abuse and corruption (which were not nearly as prevalent in the 1960s as they later became), a lack of trained and experienced entrepreneurs and people in senior and middle management and technical positions led to delays and inefficiencies in the use of resources in all sectors: the states of Africa were not equal to the economic role assigned to them, and markets would in many cases have been more efficient and probably more egalitarian too.

On the other hand, who can doubt that many of the sources of these problems lay in the African experience of centralized colonial authority, colonial trading monopolies and the colonial exploitation of ethnic differences? Is it not also reasonable to maintain that the failure of the colonial powers to try to 'negotiate' an accommodation between traditional political cultures and the constitutional orders constructed at independence contributed to the pervasive lack of popular identification with Africa's

[8] One of the World Bank's Africa specialists concludes as follows: 'The expectations about Africa's development should be tempered with realism...Even after successfully implementing most of the broad elements of this agenda (i.e. the Bank's programme of structural adjustment plus infrastructure reconstruction, better government, an 'enabling environment' for private capital, etc.) per capita incomes in sub-Saharan Africa will grow at best by only 1–2 percent and the incidence of poverty will still rise both in absolute and relative terms. Dependence on foreign assistance will continue unabated. There should be no illusions. The choice is either to work hard and steadily and ensure that at least these modest gains are realised or to abandon this continent of over 500 million people to live in misery and impoverishment and allow many more Somalias to surface' (Husain, 'Structural Adjustment', op. cit., p. 170).

[9] The most commonly cited example is that of President Mobutu of Zaïre, who was estimated in the mid-1980s to have in his personal bank accounts overseas a sum roughly equivalent to his country's entire foreign debt; but his is far from being an isolated case. See Richard Sandbrook, *The Politics of Africa's Economic Stagnation* (Cambridge: Cambridge University Press, 1985), p. 91; and Jean-François Bayart, *The State in Africa: The Politics of the Belly* (London: Longman, 1992), pp. 87–88.

post-colonial states? Who does not agree that the economic structures bequeathed by colonialism — producing what they did not consume, and consuming what they did not produce — have constituted major obstacles to the development of integrated national economies after independence?

It cannot really be denied, either, that the sum of the surpluses taken out of Africa by foreign companies since independence through 'transfer pricing', predatory 'management fees', tied 'aid' and other forms of unequal exchange have significantly handicapped the region's development; or that the superpowers exacerbated (and in some cases virtually created) conflicts in Africa that might well have been resolved, or at least blunted, had the resources that were expended on military 'aid' to the participants been devoted instead to trying to help overcome their differences. And it is no less widely agreed that most of the region's public debt can never be repaid; that much of it was incurred, with the encouragement of foreign 'donors', for projects that did not expand productive capacity; and that, as in Latin America, only more so, servicing this debt now constitutes a massive drag on all further development efforts.

Thus, on the one hand, sub-Saharan Africa is caught in a steadily closing neo-Malthusian vice, while on the other, it is enmeshed in a web of mutually reinforcing pathologies that seem to block every imaginable line of escape. And what does development theory have to say about this? What sense does it make of the multiple causes of the tragedy, and what does it suggest now about the possibility of getting out of it?

For a start, it is clear that a good deal of time and energy has been misspent in debates about which of the many determinants of the tragedy were the 'prime' or decisive ones, the 'principal' as opposed to the secondary contradictions, etc., or in urging that 'external' causes were in general more important than 'internal' ones, or vice versa. These arguments were really driven by fundamental political disagreements about the goals of development, the merits of socialism versus capitalism and the stances of the superpowers. Today, it seems obvious that what we need is a general explanation in which all the determinants alluded to above, and many others, find their place in an appropriately multilevelled and historical framework.

On the other hand, development theory cannot be content with merely showing why nothing can be done. This means that while it can no longer take as given the regulated system of national economies of the post-war years in which 'development theory' was first articulated — since this has been dismantled — neither can it take as given the contemporary world of 'self-regulating' global markets. For the origins of Africa's tragedy clearly lie far back in the emergence and evolution of the world capitalist economy; and the seeming impossibility of surmounting it today is also bound up with the fact that the leading industrial states have recently chosen to abandon that system of regulation to which the global economy was subject at the time when Africa was launched into independence.[10]

To restate this point more consecutively: sub-Saharan Africa was among the last parts of the world to be fully incorporated into the global system of production and exchange. The same reasons that delayed its incorporation — above all, its climate and endemic diseases — had also isolated it from cultural and technical changes originating in Western Europe or, earlier, East Asia. At the end of the nineteenth century the European colonizers thus confronted still predominantly decentralized and mostly preliterate African economic and social systems, with very low levels of labour productivity. Converting these systems to capitalist production relations would have involved trying to police a greater degree of social convulsion than the metropoles were willing to pay for (or than the more humane colonial officers, at least, would have been willing to countenance). Consequently the basic relations of production — land ownership and the division of labour — were left largely untransformed. Mineral production expanded in enclaves based on migrant labour; but outside a few areas of white settlement African households were induced, through a mixture of coercion and inducement, to shift their labour (above all, women's labour) into the production of export commodities, without other fundamental change. Productivity gains occurred, but within the limits of household production.

So long as world demand expanded, African households were able to produce steadily growing volumes of commodities for Western markets on this basis, and part of the surpluses thus produced — whatever was not siphoned off in the form of cheap supplies for metropolitan markets, high profits for metropolitan trading companies, etc. — raised their living standards and, in particular, lowered mortality (especially infant mortality) rates, setting in motion a demographic explosion that is still continuing. When world demand contracted, however, as it did in the interwar years and again in the 1970s, the African economies were extremely vulnerable, and not just in the short run. The organization of production made it difficult for higher levels of productivity to be achieved through mechanization or other means.[11] Meanwhile, in much of Africa the pressure of population on fertile land was already quite severe, while, on the other hand, the low incomes that could be earned from household production, combined with the small populations of most of the African ex-colonies, limited the domestic markets available to manufacturing industry. As a result, population growth eventually began to outpace the growth of output — worse, for several years in the early 1980s, output actually contracted — and social and political crises rapidly grew more acute.

At independence most African governments (and not just those with 'socialist' orientations) made very serious efforts to break out of this vicious circle through state-led development plans — efforts for which they

[10] i.e. in the sense indicated in Chapter 1 above, pp. 19–25.

[11] See Chapter 6 above for a development of this point.

mostly received strong support from the 'development community' and for which today they are apt to be blamed, rather than given credit. But the vicious circle of interlocking 'causes' of the tragedy listed above, stemming from the interaction of their weak agrarian production systems (and still predominantly precapitalist social structures and cultures) with the political and economic forces of the external world, was already in motion, and eventually brought both 'socialist' and capitalist development dreams relatively impartially to naught.

By the end of the 1980s the economies of much of sub-Saharan Africa could no longer even ensure the survival of a growing proportion of their inhabitants. World demand for what the region produces is, by and large, growing slowly or even declining, while world supplies are being constantly expanded; and many of the agricultural commodities in question are increasingly being produced several times more efficiently outside Africa under capitalist conditions of production, forcing prices steadily downwards towards levels at which Africans will no longer be able to live on what they can get from a day's labour producing them.[12] In Andre Gorz's arresting metaphor, more and more Africans are becoming 'supernumeraries' of the human race.

This is not, of course, the first time that a region of the world has been relegated to the margins of the global accumulation process, even if it is the first time this has happened to a whole subcontinent. Until recently, however, such regions (like, at different times, the Scottish Highlands, Ireland or the Italian south) could often at least export their labour power through emigration, but this option is not open to most Africans today.[13] In sub-Saharan Africa, at least, the only large-scale emigration that is occurring is the flow of refugees to internationally operated feeding centres in other African countries.

Meanwhile, on the other hand, the 'external' world to which Africa was and is exposed has itself been undergoing a fateful transformation. During the Bretton Woods era of national economies protected by effective international controls over capital movements, the industrialized countries experienced steady economic growth, distributed the benefits with a degree of equity (however modest) between capital and labour and between town and country, invested in their infrastructure, increasingly recognized and assisted disadvantaged groups and pursued all sorts of other social and cultural objectives, from gender equality to care for the environment, even if such goals were only very imperfectly attained; and they were also willing

[12] In neo-liberal terms, the equilibrium wage level in many African countries is increasingly likely to be below the level of subsistence. Manfred Bienefeld has noted the analogy with Nassau Senior's remark that the famine in Ireland would not kill more than a million people, and that would 'scarcely be enough to do much good' (Manfred Bienefeld, 'The International Context for National Development Strategies', in M.A. Bienefeld and Martin Godfrey (eds), *The Struggle for Development* (Chichester: Wheatsheaf, 1982), p. 30).

[13] Except for the diaspora of highly educated Africans, estimated at about half a million in the mid-1990s.

to devote, by 1960–61, over 0.5% of their expanding national incomes to official development assistance to underdeveloped regions and to set themselves the admittedly modest but none the less higher target of 0.7% for the second 'Development Decade'.[14]

But by the early 1990s these countries were adjusting to a world in which capital is free to move to wherever its total production costs are lowest, while at the same time rising productivity is continually reducing the demand for labour and lowering trade union bargaining power. The general result (most marked in the UK and the USA) has been declining real wages for the least skilled, dramatic increases in inequality of incomes and wealth, higher unemployment and declining levels of collective provision across a whole range of goals that only yesterday had defined the kind of societies people wanted to live in (such as support for gender and other kinds of equality, and the steady improvement of social services and social security) with corresponding increases in social distress, marginalization, racism, crime, violence and political alienation. This is the context in which the average share of GDP now devoted to official development assistance by the Organization for Economic Co-operation and Development (OECD) countries has *fallen* (by a third) to only 0.34%;[15] and this is the context in which, instead, the citizens of these countries are gradually becoming inured to the return in Africa and elsewhere of endemic starvation, cruelty and violence on a scale that would have been considered unimaginable, as well as completely unacceptable, twenty years ago.

The conclusion to which these considerations point is as simple as it is daunting: i.e. that, for all countries in the world, recapturing control over their own destinies requires the re-establishment of social control over capital and the resubordination of markets to social purposes. For the weakest regions, such as most of sub-Saharan Africa, this is literally a matter of life and death. For the citizens of industrialized countries, there is still room to soften and redistribute the costs of the competitive reduction of real wages on which they are embarked, although the weakest sections of these societies are already bearing most of the costs through unemployment, intermittent or part-time work, lower real wages and the contraction of social services and social security. But in a global economy without full employment the process has no convenient end-point. Half a century ago Karl Polanyi wrote that:

> To allow the market mechanism to be the sole director of the fate of human beings and their natural environment...would result in the demolition of society...Robbed of the protective power of cultural institutions [i.e. social and political institutions], human beings would perish from the effects of social exposure; they would die as victims of acute social dislocation through

[14] OECD, 'A History of the Development Assistance Committee and the Development Co-operation Directorate in Dates, Names and Figures', *OECD Working Papers* 2/38, 1994, p. 48.

[15] With the USA notably leading the decline, at only 0.2% of its GDP.

> vice, perversion, crime, and starvation. Nature would be reduced to its elements, neighbourhoods and landscapes defiled, rivers polluted, military safety jeopardised, the power to produce food and raw materials destroyed.[16]

Polanyi was convinced that the depression of the 1930s had permanently instilled this truth in the collective wisdom of the industrialized countries, but he was wrong; it has been deliberately attacked and discredited by those who thought they stood to gain from a return to a market-driven world, and what we see in Africa are some of the first victims. If development theory is to remain a useful intellectual endeavour today, it must surely start out from an explicit engagement with this issue and the fundamental and long-term political questions it raises.

The new global order, which will, surely, have to be rebuilt yet again in the next half century, will not be a replica of the Bretton Woods system. To the extent that a much more open global market for goods and capital (but not labour) is accepted, the new system will require new institutions of global government, which will have to be more representative of, and accountable to, people (as opposed to wealth) than most global institutions are now. But, in a world in which the market is once more subordinated to social purposes, rather than the other way round, it will be possible to envisage bringing the tragedy in Africa to a halt and resuming a process of development. Writing off Africa's debt, already widely accepted on both moral and economic grounds, would become practical politics. The replacement of 'development assistance' lending by automatic transfers from a progressive international income tax (already proposed by the UN Development Programme) would appear as a rational form of insurance against world disorder, not as charity, as would special access for African exports to overseas markets.

In this radically changed external context, the current *de facto* recolonization of Africa by aid consortia, World Bank structural adjustment teams, the UN High Commission for Refugees, the UN Food Programme and a host of other agencies could also be transformed: it could give way to a new long-term, open and accountable system of collaboration between domestic and supranational political leaders and public servants, based on principles of mutual interest in creating the economic and social preconditions for a new and more genuine sovereignty. Among these preconditions, it must be recognized, will be some to which the neo-liberal critique of past policies and the new institutionalist analysis of old institutions have made valid contributions; in some sectors more scope will need to be given to markets, in others more attention will have to be paid to the construction of non-market institutions that are resistant to politicization and corruption and yield socially efficient results. On the other hand, the accumulating evidence — from Japan and the NICs of East Asia, to China and 'Rhineland capitalism' — is also clear: for most countries, and certainly

[16] Karl Polanyi, *The Great Transformation* (Boston: Beacon Press, 1944), p. 73.

most 'latecomers' to industrialization, national success in the global marketplace depends on coherent long-term strategic action by states, and the construction and maintenance of a dense web of 'intermediate' institutions (banks, financial and technical services, training, and infrastructure of all kinds) that the market needs but does not itself provide. Yet the already critical weakness of Africa's states and institutional structures has been severely aggravated in the era of 'structural adjustment', in some cases to the point of collapse. Confronting this issue, and the fundamental political choices it presents, is surely a precondition for any new development theory that will be worth taking seriously.

It will be objected that the problem is much more complex than this, and this is true; there is certainly no assurance that any given African country will find its way out of the complex of vicious circles in which it has become trapped, even in a global environment that holds out some genuine encouragement. But there is also, as I have already remarked, a discourse of 'complexity' in which everything is dissolved into its details, and the possibility of abstracting and trying to act on the main elements and forces at work in the world is obscured (if not actually denied); and for all its shortcomings the great merit of development theory has always consisted in being committed to the idea that we can and should try to change the world, not just contemplate it — which means, in practice, being willing to abstract from the detail, to identify structures and causal relationships and to propose ways of modifying them.

For development theory, then, the question is not whether the above interpretation of the African tragedy is short on detail, but whether a closer study of its details changes its broad outlines, and if so how. More generally, what the African tragedy makes concretely clear is that development theory can no longer confine itself to the 'Third World' (or any part of it), or treat the 'developed world', and whatever global arrangements it makes, as a 'given' (benign or otherwise). 'Development studies' can no longer be conceived of as a kind of area studies, and the politics of development theory and development studies must become more explicit than ever before.

Index